TECHNICAL EVALUATION REPORT

EVALUATION FINDINGS OF THE LTL 2000 PAVEMENT MARKING RETROREFLECTOMETER

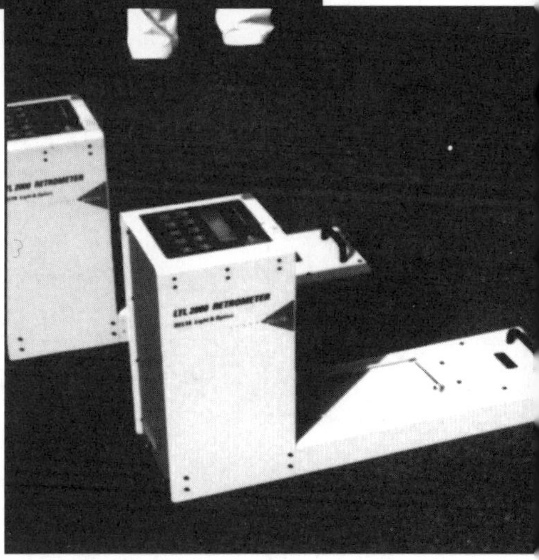

HITEC

Prepared by the
Highway Innovative Technology
Evaluation Center (HITEC), a service
center of the Civil Engineering
Research Foundation (CERF)

CERF REPORT: #40469
January 2000

Abstract

The Highway Innovative Technology Evaluation Center (HITEC) serves as a clearinghouse for implementing highway innovation by providing nationally-focused, collaborative evaluations of new products and technologies.

This report, *Evaluation Findings of the LTL 2000 Pavement Marking Retroreflectometer*, presents the results of a detailed evaluation for one out of six pavement marking 30-meter retroreflectometers that were evaluated in this program. The evaluations were designed to test the measurement bias, repeatability, and reproducibility of handheld and mobile retroreflectometers produced by several manufacturers.

Library of Congress Cataloguing-in-Publication Data

Evaluation findings of the LTL 2000 Pavement Marking Retroreflectometer / prepared for the Highway Innovative Technology Evaluation Center (HITEC); prepared by the Texas Transportation Institute, the Texas A&M University System, College Station, TX.

P. cm. -- (Technical Evaluation Report)

"A service center of the Civil Engineering Research Foundation (CERF)"

"December 1999"

ISBN 0-7844-0469-0

1. Reflectors (safety devices) -- Testing -- Equipment and supplies.
2. Reflectometer -- Testing.

I. Highway Innovative Technology Evaluation Center (U.S.)
II. Civil Engineering Research Foundation.
III. Texas Transportation Institute.
IV. Series.

TE228 .E936 1999

625.7'94--dc21

99-058015

Library of Congress Catalog Card No: 99-058015
ISBN 0-7844-0469-0
Manufactured in the United States of America.

Acknowledgments

The Highway Innovative Technology Evaluation Center (HITEC), a service center of the Civil Engineering Research Foundation (CERF), prepared this report. HITEC wishes to acknowledge the special contributions of individuals whose efforts and suggestions have significantly influenced this report. Notable, this report is based on work by a technical evaluation panel who volunteered to draft the evaluation plan for this project and carry out its objectives. The panel is composed of: Chairman Thomas Neal, Jr., KTA-TATOR, Inc.; John Arens, FHWA; Jim Carlson, Minnesota DOT; Daniel Centa, City of Pueblo DOT; Mike Crow, Kansas DOT; John Geisel, Pennsylvania DOT; Gene Hanson, Arizona DOT; Peter Hatzi, FHWA; Gene Hawkins, Texas Transportation Institute (former panelist); Neil Hodson, 3M; Alan Lafferty, Florida DOT; Rhonda Lewis, Delaware DOT; Bruce Littleton, Delaware DOT; Jimmy Lynch, North Carolina DOT; James Migletz, Graham-Migletz Enterprises, Inc.; C. Cameron Miller, National Institute of Standards and Technology (NIST); Jay Rennilson, Rennilson Consulting Services (RCS); Judy Shelato, Indiana DOT (former panelist); Jo Anne Tingle, Kentucky Transportation Cabinet; Frank Vukovits, Indiana DOT; and Brian Zimmerman, Michigan DOT.

I wish to thank the staff of the Texas Transportation Institute who served as a consultant and were responsible for producing this report. That staff included Gene Hawkins, Paul Carlson, and Mike Fontaine. Mike Fontaine did an excellent job in writing the majority of this report. Johan Bemelen of Bemelen Engineering served as a subcontractor and was responsible for developing the evaluation plan.

I would also like to extend a special thank you to Jimmy Lynch, Jim W. Kellenberger, and the North Carolina DOT who graciously volunteered their highways for the field testing, including all the necessary traffic control.

John Arens at the FHWA Turner Fairbank Research Center and Piotr Domanski and W. Vance Payne at NIST should also be acknowledged for opening their laboratories to accommodate the laboratory testing. Gene Ingwerson of the Kansas DOT also provided help during the laboratory testing. I would also like to acknowledge Neil Hodson and 3M for their assistance in the evaluation. They provided the test panels for the lab testing and the results of their own study.

Among the staff who worked on this project, I wish to acknowledge the special efforts of HITEC Director and CERF Vice President, J. Peter Kissinger; Mike Higgins, HITEC Project Manager; Randy Perrin, Research Engineer; and David Ellsworth, HITEC intern; who contributed significantly to the execution of this group of reports. Finally, the assistance of CERF Project Manager, Nicole Testa in managing the production of this report is appreciated.

Publication of this report is made possible, in part, through the contributions by members of CERF's New Century Partnership:

- Black & Veatch
- CH2M Hill Ltd.
- Charles Pankow Builders
- Charles J. Pankow Matching Grant
- Kenneth A. Roe Memorial Program
- Lester B. Knight & Associates, Inc.
- Parsons Brinckerhoff, Inc.
- The Turner Corporation

Harvey M. Bernstein

Harvey M. Bernstein
President, Civil Engineering Research Foundation

Disclaimer

This document is based on work supported by the Federal Highway Administration under Cooperative Agreement No. DTFH61-93-X-00011.

Any opinions, findings, and conclusions or recommendations expressed in this publication are those of the Highway Innovative Technology Center and do not necessarily reflect the view of the Federal Highway Administration.

This report is the result of an impartial, consensus-based approach to evaluating innovative highway technology in accordance with the HITEC Technical Protocol. The data presented are believed accurate and the analyses credible. The statements made and conclusions drawn regarding the product evaluated do not, however, amount to an endorsement or approval of the product in general or for any particular application.

Preface

The U.S. Congress, in 1992, required that the Manual of Uniform Traffic Control Devices (MUTCD) set minimum requirements for the retroreflectivity of highway pavement markings. This requirement generated a strong interest in retroreflectometers, instruments that measure the nighttime visibility of such markings. To ensure that all retroreflectometers meet the requirements of the state highway agencies, FHWA initiated a standard HITEC group evaluation of the devices.

Historically, the introduction of new or innovative technologies, such as retroreflectometers, to the highway community required that a new product be demonstrated to many, if not all state highway agencies. This inefficient, time consuming, and often-costly practice, prohibited many small companies and entrepreneurs from attempting to introduce innovation into the marketplace. To overcome these barriers, HITEC was established in 1994 in cooperation with FHWA, the American Association of State Highway and Transportation Officials (AASHTO), and the Transportation Research Board (TRB) to accelerate the process of introducing technological advances into the highway community.

HITEC is structured to facilitate the conduct of consensus-based, nationally accepted performance evaluations of new or innovative technologies for the highway community. While the term "new or innovative technologies" connotes "high tech" products, often associated with the computer industry, HITEC is available to evaluate almost any product, system, service, material, equipment or other technology that the owner believes can be used on the nation's highways.

As mentioned above, HITEC incorporates a consensus-based approach to its performance-based evaluation process. The key feature of the HITEC process is the formation of the Technical Evaluation Panel, which is composed of key representatives from the user community, academia, and the private sector. The Technical Evaluation Panel, with the cooperation and assistance of the applicant(s), identifies the specific issues and concerns requiring resolution for these products to be adopted by the highway community. These evaluation guidelines are compiled and documented, serving as the testing protocol for the subsequent evaluation. The Evaluation Plan for this group evaluation can be downloaded from the HITEC web site at www.cerf.org/hitec. As with every HITEC evaluation, the Technical Evaluation Panel oversaw the development and execution of the Evaluation Plan, and ultimately, reviewed and approved the following evaluation findings.

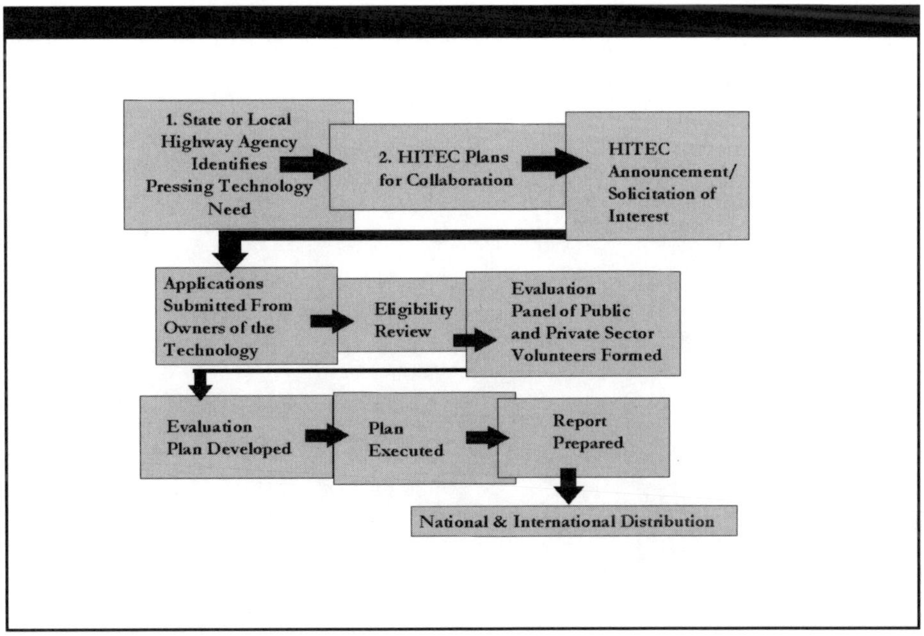

This publication is one of eight reports for the HITEC group evaluation of pavement marking retroreflectometers:

- This report, *Evaluation Findings of the LTL 2000 Pavement Marking Retroreflectometer,* provides the results of the testing called for in the HITEC Evaluation Plan and measures the performance of the retroreflectometer against the criteria in the Evaluation Plan.

- Five additional reports will be published (one for each retroreflectometer that participated in this evaluation) that provide the individual results for each device.

- A final, comprehensive summary report will be published once all the individual reports have been released and the information and data compiled.

- The HITEC Evaluation Plan that fully describes the scope and details of the HITEC test program is only available on the HITEC web site at www.cerf.org/hitec.

The individual and summary reports can be ordered from ASCE publications at 1.800.548.2723 or 703.295.6300. If you are not an ASCE member and live in Europe, please call 44-1462-437-9333. Alternatively, the reports can be downloaded on our website at www.cerf.org/hitec.

Contents

Technical Evaluation Panel Key Contacts

Product: **LTL 2000**

Chair: **Thomas Neal, Jr.,**
KTA-TATOR, Inc.

Panelists:

John Arens
Federal Highway Administration

Daniel Centa, P.E.
City of Pueblo Department of Transportation

Mike Crow
Kansas Department of Transportation

John Geisel
Pennsylvania Department of Transportation

Peter Hatzi
Federal Highway Administration

Neil Hodson
3M Corporation

Alan Lafferty
Florida Department of Transportation

Rhonda Lewis
Delaware Department of Transportation

Bruce Littleton
Delaware Department of Transportation

Jimmy Lynch, P.E.
North Carolina Department of Transportation

James Migletz
Graham-Migletz Enterprises, Inc.

C. Cameron Miller, Ph.D.
National Institute of Standards and Technology

Jay Rennilson
Rennilson Consulting Services (RCS)

Jo Anne Tingle, P.E.
Kentucky Transportation Cabinet

Frank Vukovits, P.E.
Indiana Department of Transportation

Brian Zimmerman
Michigan Department of Transportation

Former Panel Members:

Jim Carlson
Minnesota Department of Transportation

Gene Hawkins Ph.D., P.E.
Division Head
Texas Transportation Institute

Judy Shelato
Indiana Department of Transportation

HITEC Staff: **Michael Higgins, P.E.**
Randy Perrin

Consultants:
Texas Transportation Institute
Gene Hawkins, Ph.D., P.E.
Paul Carlson, P.E.
Mike Fontaine

Client:
Fritz Wang
Flint Trading , Inc.
P.O. Box 160
Thomasville, NC 27360
Phone: (336) 475-6600
E-mail: flintti@compuserve.com
Web site: www.flinttrading.com

Subcontractor:
Bemelen Engineering Consultants, Inc.
Johan Bemelen
Principal

CHAPTER 1

Introduction

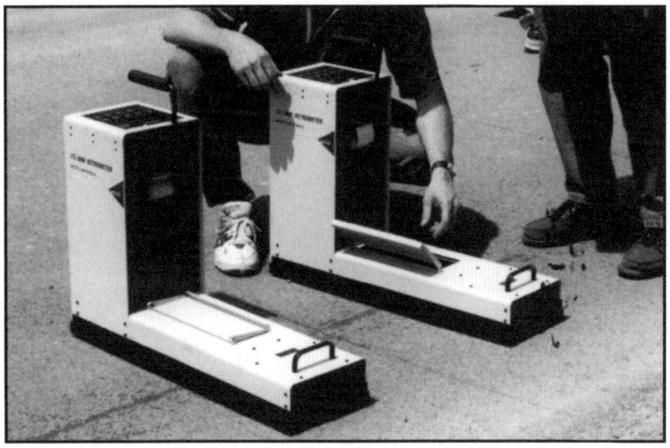

LTL 2000 Pavement Marking Retroreflectometer

Pavement markings are a critical element of the roadway transportation system. Because they are located directly within the road user's cone-of-vision, they are positioned to provide continuous information that helps the road user to correctly position the vehicle on the roadway. They can also be used to supplement the information provided by traffic signs and signals. Pavement markings are retroreflectorized so that they can function in both daytime and nighttime conditions. The retroreflective properties of pavement markings allow light from vehicles' headlights to be redirected back to drivers, enabling them to see the markings in nighttime conditions.

The retroreflective characteristic of pavement markings has assumed greater importance to transportation agencies as they have recognized the value of pavement markings as a traffic control device. Many agencies are now measuring the retroreflectivity of pavement markings when they are applied to ensure the markings meet specifications. In recent years, the Federal Highway Administration (FHWA) has been conducting research on pavement marking retroreflectivity to determine the levels of retroreflectivity needed by road users. This research was partly prompted by the Department of Transportation and Related Agencies Appropriation Act of 1992. In this act, Congress required the DOT to revise the *Manual on Uniform Traffic Control Devices* (MUTCD) to include "*a standard for a minimum level of retroreflectivity that must be maintained for pavement markings and signs which apply to all roads open to public travel.*" These minimum levels of retroreflectivity would define end-of-service life values for existing pavement markings. The FHWA is currently in the process of determining how to best implement these minimum levels of retroreflectivity. One possibility is that agencies will be required to measure pavement marking retroreflectivity on regular intervals.

A complicating factor in the measurement of pavement marking retroreflectivity is the adoption of a new measurement geometry in the United States. In order to provide for consistent measurement of retroreflective properties of pavement markings, the American Society of Testing and Materials (ASTM) has established standard measurement geometries for pavement markings. Variables in the standard include the entrance angle and the observation angle. Figure 1 illustrates these angles. For many years, U.S. transportation agencies used entrance and observation angles that represented a 12 or 15 meter viewing geometry. More recently, ASTM has adopted the 30 meter geometry as the standard for measuring pavement marking retroreflectivity. This geometry simulates the performance of a marking that is located 30 meters in front of a vehicle. It is also the same geometry used by the European Committee on Standardization (CEN). There is no direct relationship between earlier geometries and the 30 meter retroreflectivity measurements. As a result, 12 and 15 meter instruments cannot be used to measure pavement marking retroreflectivity at the 30 meter geometry in a manner that complies with the ASTM standard. In addition, the FHWA research on pavement marking retroreflectivity used the 30 meter geometry.

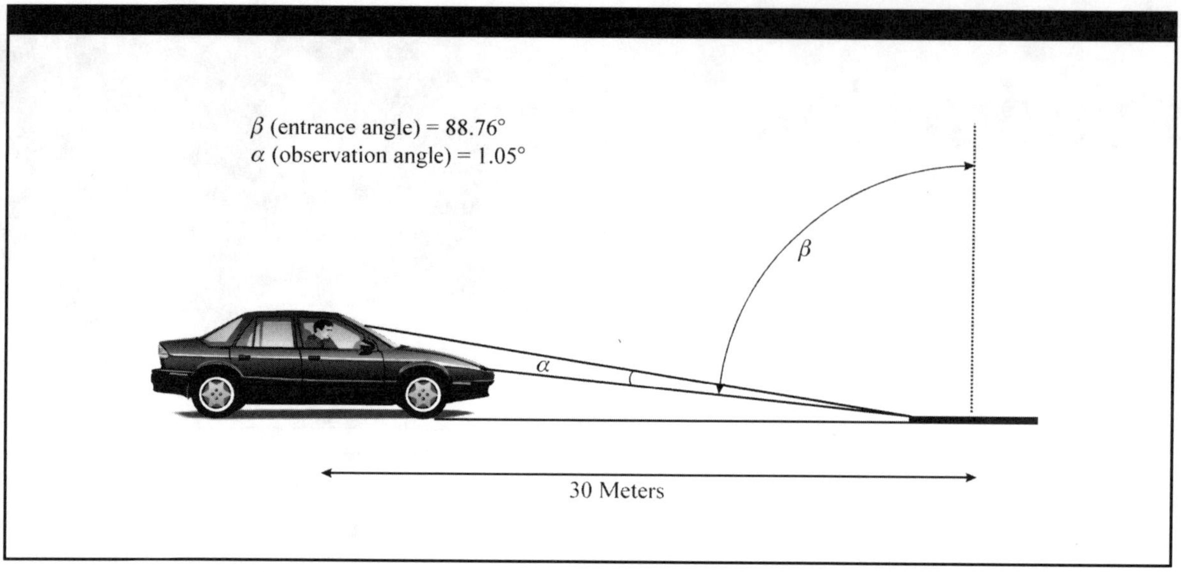

Figure 1. 30 meter Geometry for Pavement Marking Retroreflectometers

The increasing emphasis on measuring pavement marking retroreflectivity, combined with the change in measurement geometry, is creating an expanding market for instruments that measure pavement marking retroreflectivity using the 30 meter geometry. To help U.S. transportation agencies determine which instrument would best suit their needs, the Highway Innovative Technologies Evaluation Center (HITEC) conducted an evaluation of 30 meter geometry pavement marking retroreflectometers.

Both 3M and the American Association of State Highway and Transportation Officials (AASHTO) studied the performance of handheld retroreflectometers. The study conducted by Neil Hodson for 3M evaluated four types of portable pavement marking retroreflectometers using a set of target panels[1]. The reflectometers were selected at random from state departments of transportation, and each agency did their own measurements of the panel sets. The results of this study showed that there was considerable measurement variability between the different types of retroreflectometers and within each instrument type. Also, the study showed that repeatability and reproducibility were greater at lower levels of retroreflectance. Another finding of this evaluation was that photometric laboratories tended to produce larger retroreflectivity values than portable retroreflectometers. An AASHTO NTPEP study compared three portable retroreflectometers, one of which used a 15 meter geometry[2]. These retroreflectometers were used to measure materials on the NTPEP test decks. Results showed that there were fairly good correlations between the reflectometers for an individual material and location, but measurements made with one retroreflectometer could not be accurately used to predict values using another retroreflectometer.

1.1 Objectives of Evaluation

While earlier studies provided some initial information on the performance of 30 meter retroreflectometers, more research was needed. The HITEC evaluation of pavement marking retroreflectometers was conducted to meet the following objectives:

- Evaluate and document the performance of handheld and mobile retroreflectometers,
- Provide input that will aid in the development and refinement of standards and specifications for retroreflectometers, and
- Provide sufficient information to help potential users make purchasing decisions.

1.2 Scope of Evaluation

All vendors with the intention of marketing pavement retroreflectometers in the U.S. were invited to participate in the HITEC evaluation. Four handheld and two mobile technologies agreed to participate and completed the evalua-

1 Hodson, N. "Comparison of Pavement marking Portable Retroreflectometers", 3M Corporation, St. Paul, MN (1999).

2 American Association of State Highway and Transportation Officials, *National Transportation Product Evaluation Program Report 123: Comparison of Portable Retroreflectometers, Washington, D.C. (1997).*

Type of Retroreflectometer	Manufacturer/Vendor	Model
Handheld	Advanced Retro Technologies (ART)	MX-30
	Mirolux Products, Inc.	Mirolux Plus MP-30
	Mechatronic	FRT01
	Flint Trading, Inc.	LTL 2000
Mobile	Roadware	Laserlux
	Traffic Safety Systems	Ecodyn

Table 1. Technologies Evaluated

tion, as described in Table 1. This report describes the evaluation results for the LTL 2000 technology. The other technologies are described in separate reports, making a total of six individual reports. A seventh report summarizes the findings for all of the devices.

1.3 Evaluation Procedure

The evaluation of pavement marking retroreflectometers was conducted to determine how well individual technologies measured retroreflectivity with respect to the following performance characteristics:

- Retroreflectivity measurement:
  - Measurement bias,
  - Repeatability, and
  - Reproducibility;
- Physical characteristics;
- Operational procedures;
  - Environmental constraints;
- Economic issues; and
- Service requirements.

The evaluation plan consisted of three major elements: instrument information, laboratory data collection, and field data collection. These are described below.

- **Instrument information**
  General information about the instrument including factors that are not measurement-related. This information is described in Chapter 2.

- **Laboratory data collection**
  The instruments were used to measure retroreflectivity of 24 different panels. The 24 panels consisted of 3 sets of 8 panels each. The 8 panels in each set represented various colors and levels of retroreflectivity. Measurements were made under controlled and ambient environmental conditions. The evaluation procedure is described in Chapter 3 and the results are described in Chapter 4.

- **Field data collection**
  The instruments were used to measure retroreflectivity of pavement markings on actual highways. The evaluation procedure is described in Chapter 3 and the results are described in Chapter 4.

The three primary criteria used to evaluate the performance of the retroreflectometers were measurement bias, repeatability, and reproducibility. These characteristics are defined below.

- **Measurement bias**
  Since there is no accepted standard for retroreflectance, there is no absolute way to determine how accurately an instrument measures a sample. In this study, test panels were measured at several photometric ranges using a variety of techniques. The average of these measurements was assumed to be a best value for the panel, and was used as a baseline to compare the magnitudes of the instrument readings.

- **Repeatability**
  Repeatability is the ability of an instrument to obtain identical readings at the same exact point. Repeatability shows whether the unit produces consistent readings when the conditions of the measurement are unchanged.

- **Reproducibility**
  Reproducibility is the ability of devices to produce identical readings at relatively the same point. Reproducibility is examined to determine if different units produce similar results at approximately the same point.

CHAPTER 2

Instrument Description

LTL 2000 Pavement Marking Retroreflectometer

The LTL 2000 is a handheld pavement marking retroreflectometer that is manufactured by Delta Light & Optics and distributed by Flint Trading, Inc. The LTL 2000 uses the 30 meter geometry, and meets ASTM E 1710 and CEN EN 1436. Figure 2 shows two LTL 2000 retroreflectometers. The information included in this chapter was provided by the vendor and was not independently verified.

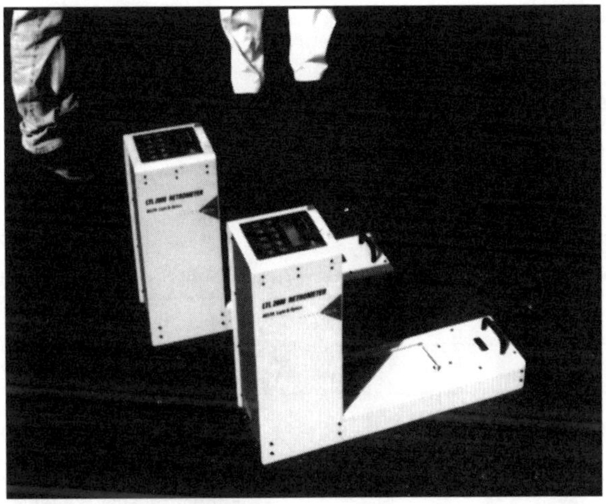

Figure 2. LTL 2000

2.1 Physical and Measurement Characteristics

Table 2 summarizes some of the important characteristics of the LTL 2000. The LTL 2000 comes with the following accessories standard:

- Carrying case
- Traceable calibration standard
- Field calibration standard
- Users manual
- Extra printer paper
- 120VAC battery charger
- Extra fuses
- Torx wrench
- Users quick guide
- Extra field calibration standard labels

2.2 Operational Characteristics

Only one person is required to operate the instrument. No special training or certification is required, and initial hands-on training is offered free of charge. The LTL 2000 can be set up by a proficient user in less than 45 seconds. This is the time it takes to move the unit from its carrying case, calibrate it, and place it on a pavement marking. The actual time to collect a reading is less than 2 seconds.

Calibration is performed in two steps. First, zero is set for the instrument, and then the actual calibration is performed. Both of these operations are initiated by the user, but controlled by the instrument. Calibration is normally done to a field standard that in turn is calibrated to a traceable reference standard. Both the traceable calibration standard and field calibration standard supplied with the LTL 2000 are made of ceramic diffuse reflector. Calibration should be performed once every two hours of continuous use, or just before taking measurements if the unit has been turned off for 30 minutes or more.

Data is stored automatically by the LTL 2000. Each data point includes the date, time, error status, retroreflectivity value, and an ID label. The data is stored in non-volatile memory and has a capacity of over 1000 measurements.

Data retrieval is accomplished via an RS-232 serial port connected to an IBM-compatible computer. Data can be retrieved using any communication program that has a terminal mode.

Data can also be printed using a built-in thermal printer. The printer can output the zero set, calibration, the most recent measurement, or all measurements stored in memory. User interface and data display is through a 10 key sealed keypad and a 1 inch by 3.75 inch liquid crystal display (LCD).

The vendor states that a variety of conditions should not impact the readings produced by the LTL 2000. Battery voltage should not impact the accuracy of the LTL 2000 if the calibration frequency and user manual instructions are followed. There are no special requirements for pavement markings of different shapes or types. No changes are necessary to measure pavement markings during the day vs. at night. The vendor also states that no adjustments need to be made for different pavement surfaces.

Standard(s) Met	CEN EN 1436 ASTM E1710
Weight in Use	27 lbs. (12 kg)
Dimensions (L x W x H)	27.8 in. x 7.9 in. x 18.5 in. (705 mm x 200 mm x 470 mm)
Observation Angle	1.05°
Entrance Angle	88.76°
Area Measured	1.8 inches x 7.9 inches (45 mm x 200 mm)
Light Source	Long life tungsten halogen lamp (CIE type A illuminant)
Range of Measurements	0 to 2000 mcd/m²/lux (linear over entire range)
Measure Profiled Markings?	Yes
Measure Wet Markings?	Yes
Recommended Operating Temperature	0° C to 45° C (32°F to 113°F)
Battery Type	Internal 12V, 3.0Ah sealed lead acid battery
Battery Life Before Recharging	Approx. 1000 readings
Memory Capacity	Over 1000 readings
Data Transfer Method	RS 232 serial interface

Table 2. Characteristics of the LTL 2000

The unit must be shipped in the carrying case or another adequately padded and fitted container. The shipping weight is approximately 50 lbs.

2.3 Maintenance

The LTL 2000 has a one year limited warranty on parts and labor. The estimated service life of the unit is 10-15 years. The vendor suggests that regular maintenance be performed on the unit once a year. The fee for this is currently $175. A factory re-certification of the traceable calibration standard, performed on an as-needed basis, costs approximately $800. It is recommended that the need for re-certification be determined during the annual maintenance for each instrument.

2.4 Cost

The base unit price includes all of the accessories listed earlier. Flint Trading also offers a heavy duty carrying case and an automobile 12VDC to 120VAC converter for an additional charge. Table 3 summarizes the current prices for these items. These prices are accurate as of July 1999.

Item	Cost
LTL 2000 Base Unit with Standard Accessories	$15,950.00
Car power adapter	$90.00
Heavy duty carrying case	$613.00

Table 3. Prices for the LTL 2000

CHAPTER 3

Evaluation Procedure

Testing of Technologies

A series of tests were conducted in order to determine the measurement bias, repeatability, and reproducibility of the LTL 2000. Laboratory studies were performed to determine the impact of different ambient light levels, pavement marking retroreflectivity levels, and environmental conditions on the instrument's ability to give consistent, accurate readings. A series of field tests were also conducted to examine the performance of the unit under real world conditions.

3.1 On-Site Representatives

Independent testers were not used to operate the test units. Due to time constraints, funding limitations, and the desire to control as many evaluation variables as possible, manufacturer's representatives were used to collect measurements while representatives from HITEC, the evaluation panel, or HITEC's consultant were present to observe and record measurements during all data collection activities.

3.2 Laboratory Tests

Laboratory tests were conducted at the National Institute of Standards and Technology (NIST) testing facilities in Gaithersburg, Maryland and at the FHWA Turner Fairbank Highway Research Center in McLean, Virginia from May 3-7, 1999. Flint Trading provided three instruments for the laboratory testing.

These tests were used to verify the measurement bias, repeatability, and reproducibility of the units in a controlled setting. These tests were performed under a variety of temperature and humidity conditions in an environmental chamber. Tests were also performed under ambient environmental conditions but with a series of different light levels.

The repeatability and reproducibility of an instrument are represented by the 95th percentile confidence interval as a percent of the mean of all applicable retroreflectivity readings. Confidence intervals are calculated based on the desired confidence level, the number of observations, and the standard deviation of the sample.

In this study, the 95th percentile confidence interval was used. This confidence interval represents the range of retroreflectivity measurements that a reading from the unit could be expected to fall into 95 percent of the time. When the confidence interval is divided by the mean reading value, a percentage is created that indicates how much variability from the mean reading can be expected with a 95 percent probability. Thus, low repeatability percentages indicate that the unit produces more consistent results than high percentages. For example, a reproducibility of ± 2 percent means that 95 percent of the time retroreflectivity readings from an instrument will be within 2 percent of the mean of the readings.

Figure 3. Test Panel Setup

Panels Tested

Standardized samples of yellow and white pavement markings were tested. White samples with a range of retroreflectivity from high to low were tested in order to simulate new versus worn markings. White sample panels were provided by 3M. Two yellow panels were also tested. A yellow panel with high retroreflectivity was provided by 3M and one with low retroreflectivity was manufactured by Potters Industries, Inc. Each panel was 6 inches wide by 2 feet long with a 4 inch by 2 foot stripe. Figure 3 shows an experimental layout of the test panels.

The retroreflectivity values shown in Table 4 are the average of many different readings of each panel obtained from laboratory photometric ranges. A total of eleven different trials were made on separate days. Four different methods were used to

Panel	Panel Set	Retroreflectivity ($mcd/m^2/lux$) "assigned value"	Color
B	1 2 3	60 61 62	White
C	1 2 3	116 115 111	White
D	1 2 3	274 288 280	White
E	1 2 3	707 693 679	White
F	1 2 3	1142 1134 1125	White
G	1 2 3	192 195 198	Yellow
H	1 2 3	77 88 91	Yellow

Table 4. Laboratory Test Panel Characteristics

measure the panels. Table 4 provides the overall average value obtained from these four methods for each panel. Appendix C includes information on the specific values given by each method, as well as the variation in readings produced between methods. The average value produced by the photometric ranges is termed the "assigned value" of the panel and is assumed to reasonably represent the retroreflectivity of the panel. This assigned value is used as a baseline for comparisons in measurement magnitude. The methods used to measure the panels included:

1. Measurements conducted in a photometric laboratory at a 15 meter distance using the direct method as described in ASTM D 4061. This was done using facilities at 3M.

2. Measurements obtained in the same 3M photometric laboratory at a 5 meter distance using the direct method in ASTM D 4061.

3. Readings collected by 3M using the luminance measurement method (a telephotometer and a lux meter at a 5 meter distance).

4. Measurements obtained at the Turner Fairbank Highway Research Center using the luminance measurement method. A different white calibration standard was used for each of the three trials.

An additional panel, panel A, was also tested. Panel A was a black panel with an assumed value of 0 mcd/m^2/lux. This panel was used to determine the ability of the instrument to reject stray light. The measurements performed on panel A are not included in the text of the report, but are summarized in Appendices A and B. The testing of panel A showed that stray light did not impact the measurements for the LTL 2000.

Environmental Chamber Testing

Each unit was tested at three different combinations of temperature and humidity in the NIST environmental chamber. These included a high temperature, high humidity test; a high temperature, low humidity test; and a low temperature, low humidity test. Each unit was allowed time to become acclimated to each environmental condition. While the temperature was held constant, there was some variability in the humidity in the environmental chamber. Table 5 summarizes the conditions tested.

Ambient Testing

Measurements were also collected under ambient conditions. On May 4 and May 5, readings were taken outdoors during the day. On May 4, measurements were collected outdoors at night. Data were also collected indoors with all lights off on May 5.

Condition	Typical Temperature	Minimum Humidity	Maximum Humidity
High Temperature, High Humidity	105° F (40.6° C)	69%	75%
High Temperature, Low Humidity	105° F (40.6° C)	13%	15%
Low Temperature. Low Humidity	35° F (1.7° C)	21%	23%

Table 5. Test Conditions in NIST Environmental Chamber

Condition	Test			Date
	Repeatability	Reproducibility	Shade	
High Temperature, High Humidity	✓	✓		5/3/99, AM
High Temperature, Low Humidity	✓	✓		5/4/99, AM
Low Temperature, Low Humidity	✓	✓		5/6/99, AM
Ambient Daylight, Day 1	✓	✓		5/4/99, PM
Ambient Daylight, Day 2	✓	✓	✓	5/5/99, PM
Ambient Night, Indoors	✓	✓		5/5/99, AM
Ambient Night, Outdoors	✓	✓		5/4/99, PM

Table 6. Laboratory Tests Performed

Test Schedule

Table 6 summarizes the tests that were conducted for each test condition. Repeatability and reproducibility tests were performed for all conditions, and an additional shade test was performed during the second day of ambient daylight testing. The purpose of the shade test was to determine if the unit produced different results depending on if it was located in direct sunlight or in a shaded area.

Repeatability Testing

Repeatability was tested for all conditions. For this test, a unit took six retroreflectivity measurements on each panel without moving the unit. This was repeated for all panels with every unit.

Reproducibility Testing

Reproducibility was also tested for each unit for all conditions. The testing sequence for the reproducibility of the units was as follows:

1. Take instrument 1 and take a reading in the center of each panel from panel set 1.
2. Repeat step 1 with panel set 2.
3. Repeat step 1 with panel set 3.
4. Repeat steps 1 - 3 with units 2 and 3.
5. Shut off unit 1 for 1-3 minutes, recalibrate, and repeat measurements as above to get replicate reading for each panel with unit 1.
6. Repeat step 5 with unit 2 and 3.

Shade Test

A shade test was performed under ambient daylight conditions on day 2. This test was performed to determine whether daytime readings taken in the shade were different from those taken in direct sunlight. The procedure for this test was as follows:

1. Take 3 readings with unit 1 on panel A (panel set 1) in direct sunlight.
2. Take 3 readings with unit 1 on panel A (panel set 1) in the shade.
3. Repeat steps 1 and 2 with units 2 and 3 and panel A from panel set 2.
4. Repeat steps 1-3 with panels B-H.

3.3 Field Tests

Field tests were conducted in North Carolina on US 1 and US 421 on June 2-3, 1999. Pavement markings on the US 421 test section had less retroreflectivity than those on US 1, but pavement condition was good at both sites. Measurements were taken on white edgelines, white skip lines, and yellow edgelines at both sites.

All measurements at each site were completed in one day, and traffic was not allowed on the markings between measurements. The same three instruments were used for the field tests as were used for the laboratory testing.

Divided Highway Testing

The divided highway testing was performed as follows:

1. Two measurements were taken on every third skip strip with unit 1 for the entire 1 mile length of the test section. These measurements were approximately 1 foot from the beginning of the stripe, and 1 foot from the end of the stripe. This resulted in 88 readings.
2. After unit 1 had been shut off for 1-3 minutes, it was recalibrated, and step 1 was repeated.
3. Unit 1 was recalibrated and an equal number of readings were taken at the same intervals as step 1 along the right (white) edgeline.
4. After unit 1 had been shut off for 1-3 minutes, it was recalibrated and step 3 was repeated.
5. Unit 1 was recalibrated and an equal number of readings were taken at the same interval as step 1 along the left (yellow) edgeline.
6. After unit 1 had been shut off for 1-3 minutes, it was recalibrated and step 5 was repeated.
7. Steps 1-6 were repeated for units 2 and 3.

CHAPTER 4

Evaluation Findings

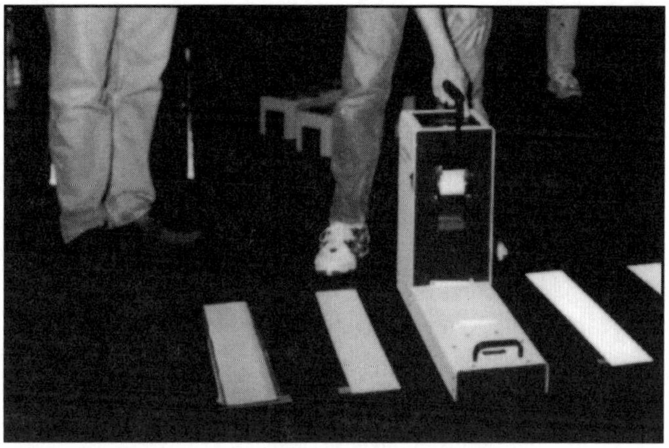

LTL 2000 Pavement Marking Retroreflectometer

The laboratory and field testing provided some useful insights into the measurement bias, repeatability, and reproducibility of the LTL 2000. Specifically, several conditions were identified that tended to impact the measurements produced by the unit.

4.1 Laboratory Testing

Table 7 summarizes the results of all testing conducted during the laboratory phase of the evaluation. These tests show that there is some variability in the measurements among the different panel sets. Appendix A offers a more detailed summary of the data collected during the laboratory testing.

Measurement	Panel							
	Panel Set	B	C	D	E	F	G	H
Minimum	All	36	80	235	606	1070	119	57
	1	45	87	252	673	1070	155	57
	2	44	90	246	633	1114	154	65
	3	36	80	235	606	1115	119	68
Mean	All	46.0	97.4	273.9	704.5	1206.2	183.1	70.4
	1	46.6	97.1	270.8	716.1	1182.8	183.1	66.3
	2	46.7	100.3	279.6	690.2	1223.7	187.9	71.2
	3	43.3	91.4	267.4	712.9	1214.4	171.9	76.9
Maximum	All	52	113	322	816	1464	219	82
	1	48	106	294	775	1304	193	70
	2	52	113	322	816	1464	219	82
	3	49	99	288	746	1298	193	82
Standard Deviation	All	1.0	5.4	13.7	32.1	60.4	11.0	4.8
	1	1.0	5.5	12.5	22.6	64.4	6.8	2.7
	2	1.2	3.3	12.5	32.6	58.3	8.6	3.0
	3	2.1	3.6	13.8	35.7	38.1	14.5	3.4
Assigned Value	Average	61.0	114.0	280.7	693.0	1133.7	195.0	85.3
	1	60	116	274	707	1142	192	77
	2	61	115	288	693	1134	195	88
	3	62	111	280	679	1125	198	91

Table 7. Panel Retroreflectivity ($mcd/m^2/lux$) Measurement Summary

Instrument and Product Variability

This evaluation attempted to identify the performance characteristics of the LTL 2000, including its ability to provide consistent readings. When measurements were taken, two factors could influence the readings: variability introduced by the instrument and variability created by the sample. While the test panels used in laboratory evaluation were made of the most uniform materials available, they were not absolutely uniform. If readings were taken on different places on the test panels, it is possible that different readings would be obtained. Thus, the results presented here may show the impact of both variations in the test panels as well as variations in the readings produced by the instrument. The impact of product variability would be most apparent in the reproducibility testing, where instruments were moved between readings. It was not possible to separate the influence of these two factors, so readers should be aware that the reproducibilities shown in this chapter are not entirely due to variation in the instrument.

Potential Sources of Error

During the high temperature, high humidity testing moisture occasionally formed on the panel surfaces. Compressed air was used to blow the moisture off the panel surface. This was only done when there was visible moisture on the panel. It is possible that condensation could have been present on the panel surface during measurements, and this may have impacted the readings produced by the instrument.

At the start of the high temperature, high humidity testing, a "High Zero Warning" error was displayed by unit 3 during the calibration process. The unit appeared to be functioning properly, and was used to collect data during the high temperature, high humidity testing. This data is included in the results presented in this report. However, it is possible that the performance of unit 3 is not representative of typical LTL 2000 performance. As a result of this error message, modifications were made to unit 3. Appendix E describes these changes. Readers should be aware that the results shown for the high temperature high humidity testing may be influenced by these two factors.

Environmental Chamber Testing

Figure 4 shows the combined mean retroreflectivity readings from the repeatability portion of the environmental chamber testing. The values shown are the mean of measurements taken over all three panel sets with all instruments during the repeatability testing.

For both high temperature conditions, the LTL 2000 produced lower retroreflectivity values than the assigned values for panels B, C, D, G, and H. The high temperature testing produced larger values for panels E and F than the assigned values. The low temperature, low humidity testing produced values that were larger than the assigned values for panels D and F.

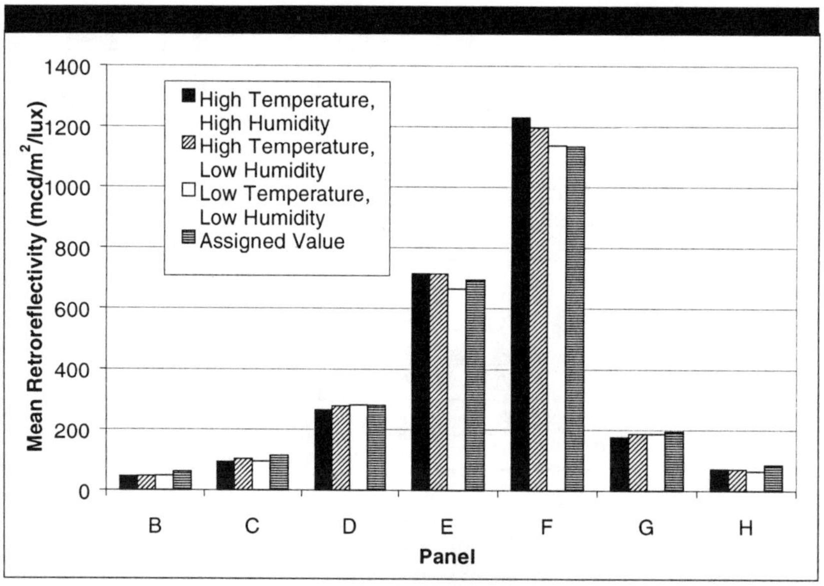

Figure 4. Mean Retroreflectivity for Environmental Chamber Testing

Figure 5 shows the repeatability of the unit under all three environmental chamber conditions. The repeatability is represented by the 95th percentile confidence interval for a panel as a percent of the mean retroreflectivity reading for that panel. Confidence intervals are calculated based on the desired confidence level, the number of observations, and the standard deviation of the sample. In this case, the 95th percentile confidence interval represents the range of retroreflectivity measurements that a single reading from the unit could be expected to fall into 95 percent of the time. When the confidence interval is divided by the average reading, a percentage is created that indicates how much variability from the mean reading can be expected with a 95 percent probability. Thus, low repeatability percentages indicate that the unit produces more consistent results than high percentages.

Repeatability is under ± 1.2 percent for panels B through H for the high temperature, low humidity and low temperature, low humidity conditions. The high temperature, high humidity condition consistently produced the largest repeatability percentages.

Figure 6 shows the reproducibility of the LTL 2000 for the environmental chamber tests. The reproducibility is also expressed by the 95th percentile confidence interval as a percentage of the mean retroreflectivity.

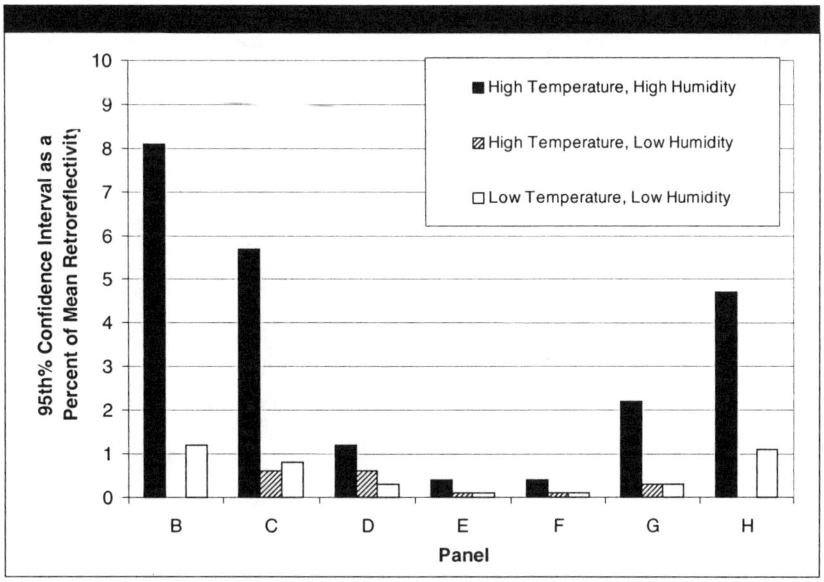

Figure 5. Repeatability for Environmental Chamber Testing

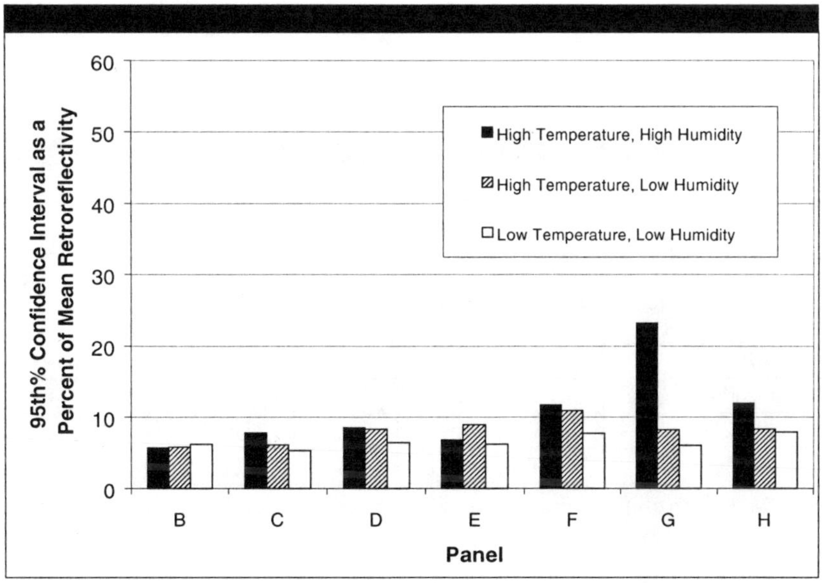

Figure 6. Reproducibility for Environmental Chamber Testing

The high temperature, high humidity condition produced the largest reproducibilities for panels C, D, F, G, and H. The low temperature, low humidity condition generated the smallest reproducibilities for all panels except for B.

Ambient Conditions Testing

Figure 7 shows the combined mean retroreflectivity readings from the repeatability portion of the ambient conditions testing. The daytime testing produced the largest readings for panels D, E, F, and H while the nighttime, outdoor testing generated the largest readings for panels B, C, and G.

The ambient condition testing produced retroreflectivity values that were lower than the assigned values for panels B, C, D, G, and H. The measurements were larger than the assigned values for panels E and F.

The retroreflectivity values measured during the repeatability testing are summarized in Table 8. These values are the mean of measurements taken over panels sets 1, 2, and 3 with all instruments.

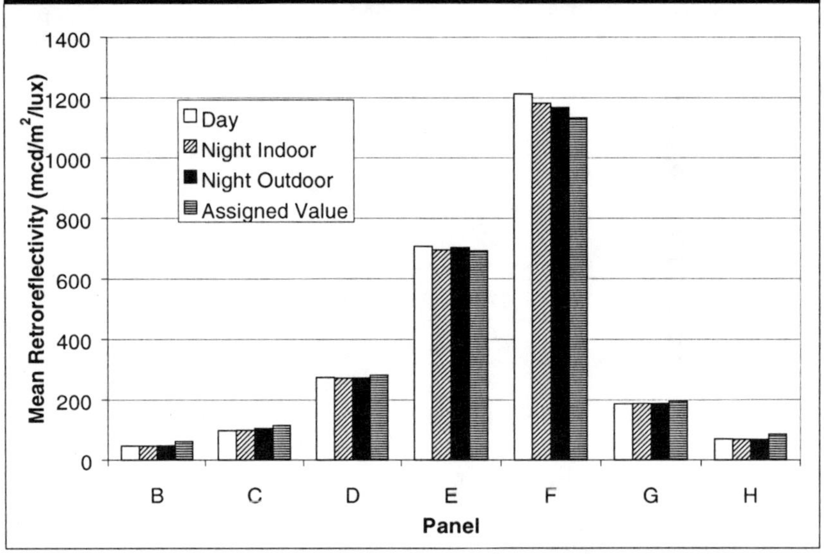

Figure 7. Different Ambient Light Levels Retroreflectivity Measurements

Condition	Panel						
	B	C	D	E	F	G	H
High Temp., High Humidity	45.4	93.6	265.3	714.4	1230.8	175.7	71.4
High Temp., Low Humidity	46.0	102.6	278.7	714.3	1196.8	187.1	70.5
Low Temp., Low Humidity	47.1	94.8	281.8	663.0	1136.5	182.9	64.2
Daytime	46.4	96.9	273.9	707.3	1212.6	185.6	69.7
Night, Indoors	45.7	97.6	271.8	695.3	1182.6	186.3	68.5
Night, Outdoors	47.5	104.5	273.1	704.2	1168.4	186.8	69.0
Assigned Values	61.0	114.0	280.7	693.0	1133.7	195.0	85.3

Table 8. Mean Panel Retroreflectivity (mcd/m²/lux) For Repeatability Tests

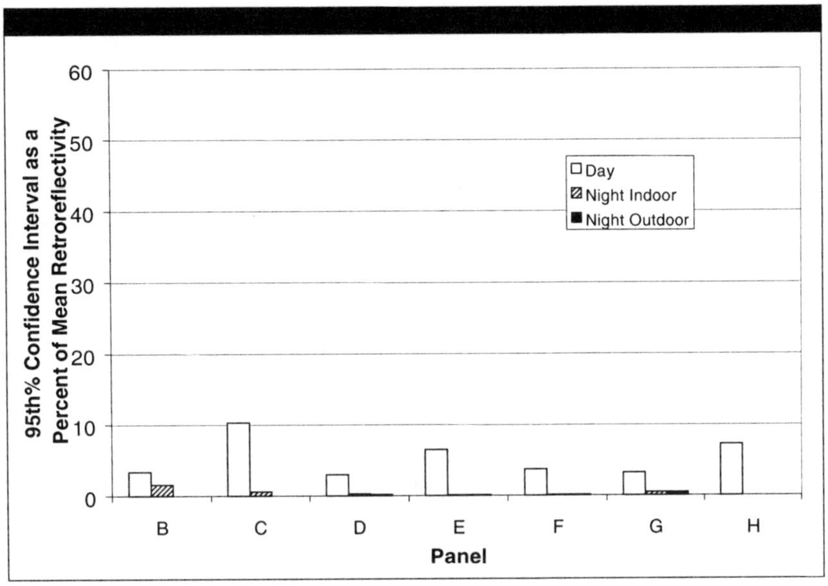

Figure 8. Repeatability for Different Ambient Light Levels

Figure 8 shows the repeatability of the unit for the different ambient light conditions. The repeatability is less than ±11 percent for the daytime testing, and less than ±2 percent for both of the nighttime tests. Daytime readings consistently had a larger repeatability than the nighttime conditions. The large repeatability percentages for daytime testing may be partially attributable to changes in the environmental conditions (amount of sunlight, temperature, humidity, etc.) during the course of the testing.

Figure 9 shows the reproducibility of the instrument for the different light conditions. The reproducibility is less than ± 12 percent for all conditions. The reproducibility for the white panels seems to be comparable to the yellow panels with similar retroreflectivities. The nighttime, outdoor results showed better reproducibility than either the daytime or nighttime indoor testing for all panels except G.

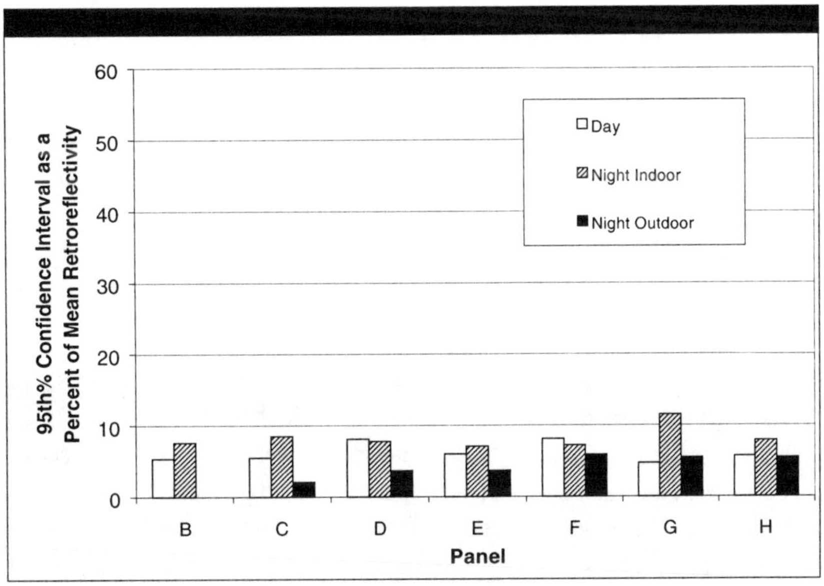

Figure 9. Reproducibility for Different Ambient Light Levels

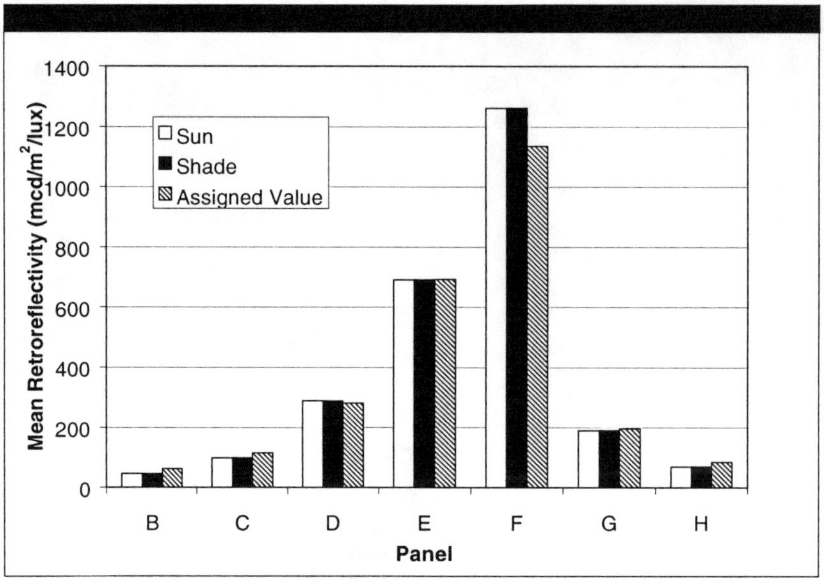

Figure 10. Shade Test Mean Retroreflectivity

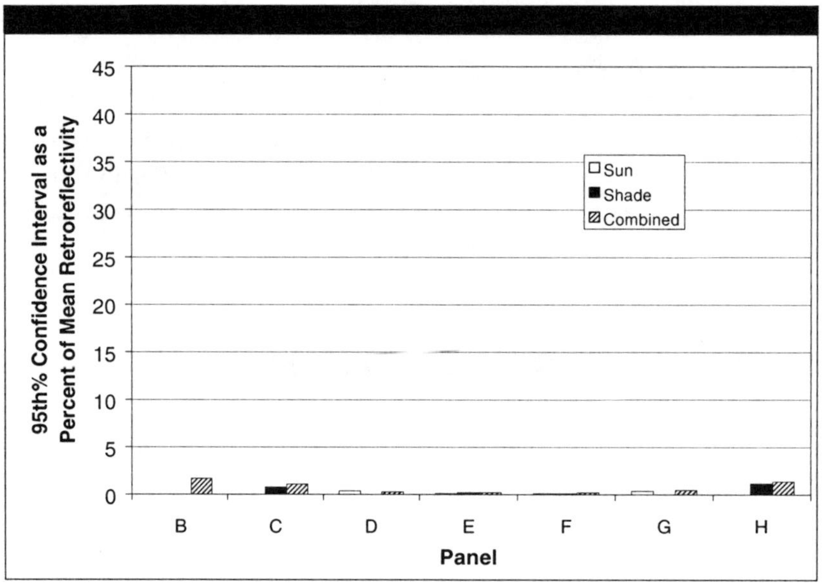

Figure 11. Sun vs. Shade Reproducibility

Shade Test

The shade test revealed that the readings taken in the shade were larger than the readings taken in the sun. However, the readings taken in the shade were never more than 1 percent larger than the readings taken in the sun. Figure 10 shows the mean retroreflectivity values obtained in the shade test.

Figure 11 shows the reproducibility for the data collected in the sun and in the shade. The combined reproducibility is less than ± 2 percent, showing a good correlation between the data collected in the sun and the data collected in the shade. The reproducibility for the yellow panels G and H appears to be similar to that for white panels with similar retroreflectivities.

4.2 Field Tests

The LTL 2000 was used to collect pavement marking retroreflectivity measurements on US 1 and US 421 in North Carolina. The white skip lines, white right edgeline, and yellow left edgeline were measured over a length of one mile at both sites. The pavement markings at the US 1 test section were less than a year old, and were in excellent condition. The markings on US 421 were not in as good a condition as the US 1 markings and appeared to possess less retroreflectivity.

Table 9 summarizes the results of the field data collection. Appendix B shows a detailed summary of the field data collected, summarized in 480 ft intervals.

Product and Instrument Variability

The field testing data was also impacted by variability in the pavement markings. By their nature, pavement markings will not have uniform retroreflectance since beads will not be dispersed uniformly. Thus, minor differences in the placement of an instrument on a marking can result in changes in the retroreflectivity readings. This variability is likely to be even higher on worn pavement markings. The reproducibility values shown in this report may be influenced by the characteristics of the markings, and may not be entirely attributable to variability in the instrument.

Site	Type of Marking	Unit	Minimum	Mean	Maximum	Standard Deviation
US 1	Skip Line	All	381	466	561	37.7
		1	394	466	546	36.1
		2	381	455	528	35.7
		3	400	478	561	37.7
US 1	White Edgeline	All	299	443	546	50.5
		1	299	441	526	49.1
		2	302	432	519	48.1
		3	326	457	546	51.2
US 1	Yellow Edgeline	All	101	269	310	24.4
		1	101	267	300	23.7
		2	104	264	298	24.4
		3	114	275	310	24.0
US 421	Skip Line	All	18	93	149	27.8
		1	18	92	145	27.7
		2	18	92	146	27.2
		3	20	96	149	28.6
US 421	White Edgeline	All	21	148	273	50.1
		1	21	147	261	49.4
		2	21	144	243	48.4
		3	21	152	273	52.5
US 421	Yellow Edgeline	All	52	83	102	7.8
		1	54	82	100	7.5
		2	52	81	101	7.5
		3	56	85	102	7.8

Table 9. Field Data Retroreflectivity Summary

White Skip Lines

The average skip line retroreflectivity values at the US 1 test site ranged from 381 mcd/m²/lux to 561 mcd/m²/lux. Retroreflectivity values for the US 421 skip lines were much lower, ranging from an average of 18 mcd/m²/lux to 149 mcd/m²/lux. Since no absolute value of the pavement marking retroreflectivity is known, it is impossible to say how accurate the units measured the markings. The three units tested seemed to produce consistent results. The retroreflectivity readings for the white skip lines are shown in Figures 12 and 13.

The reproducibility of the units was examined for the white skip lines. The three units tested each took two readings at approximately the same spot on the skip stripe. The reproducibility was determined for each of the 88 data points by determining the 95th percentile confidence interval and dividing it by the average of the six readings.

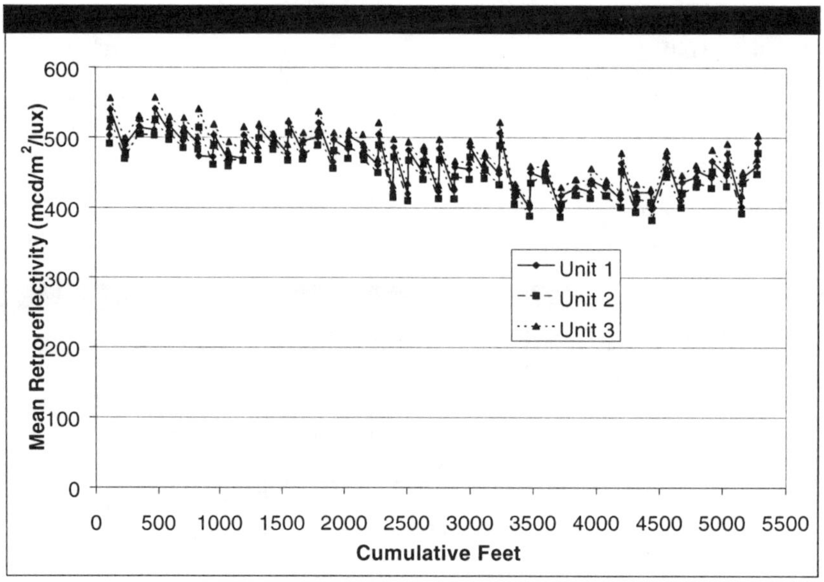

Figure 12. US 1 Skip Line Retroreflectivity

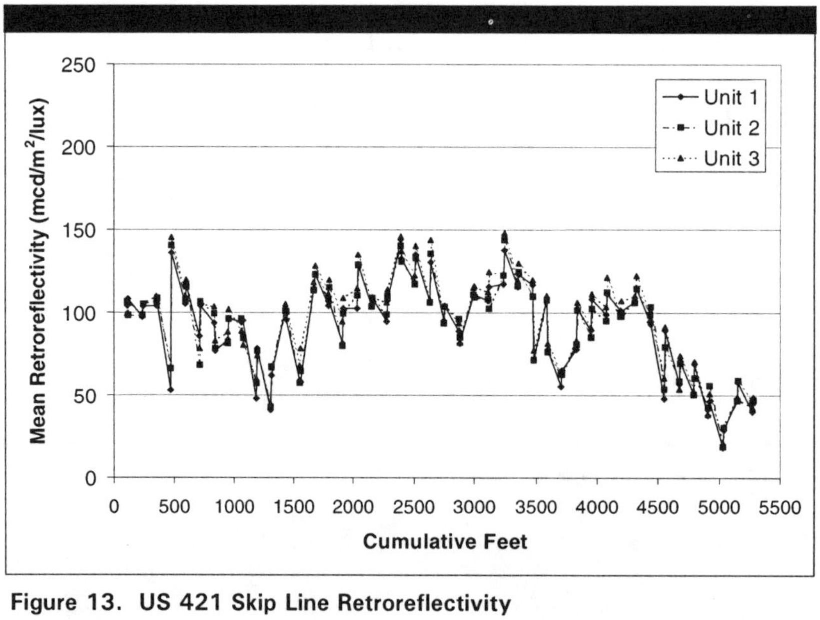

Figure 13. US 421 Skip Line Retroreflectivity

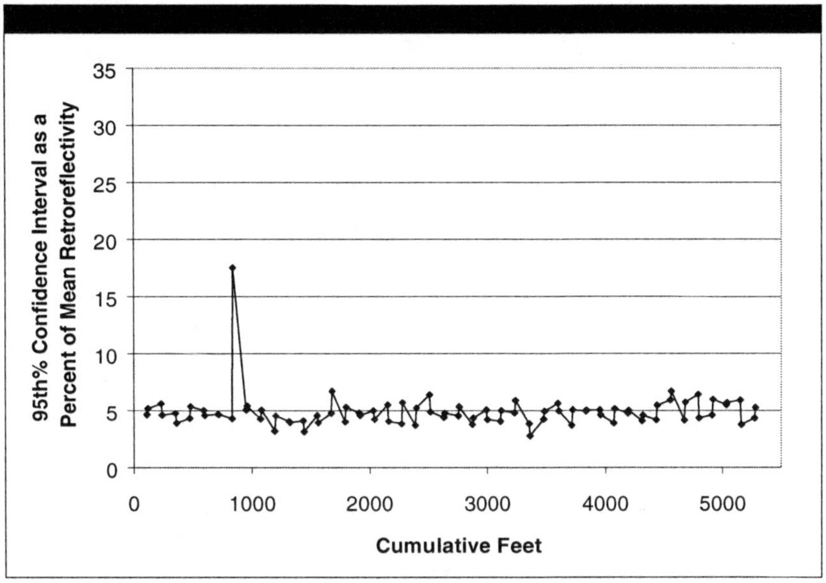

Figure 14. US 1 Skip Line Reproducibility

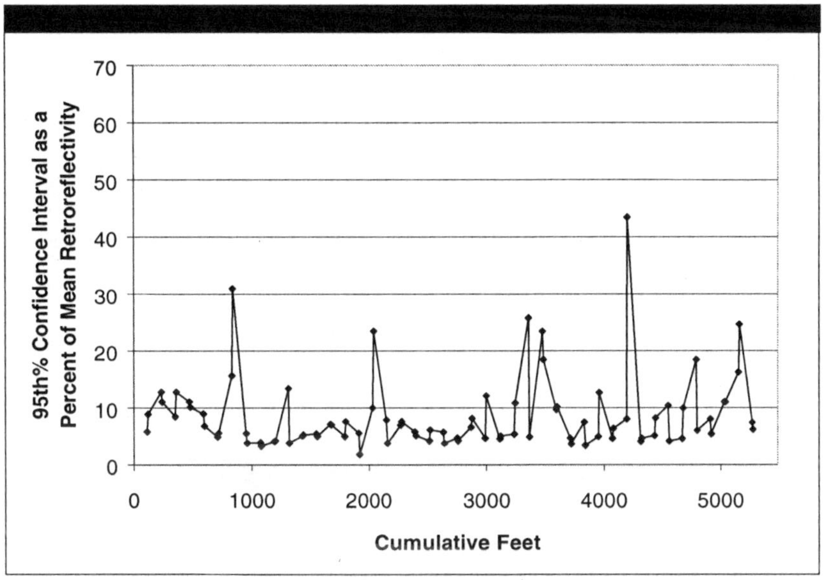

Figure 15. US 421 Skip Line Reproducibility

Figure 14 shows the reproducibility for the units at the US 1 test site. The units showed reproducibility close to that achieved for the same retroreflectivity values in the ambient light laboratory testing. The average reproducibility across all 88 data points was ± 4.9 percent. About 60 percent of the data points had a reproducibility of less than ± 5 percent, and 99 percent had a reproducibility of less than ± 10 percent.

Figure 15 shows the reproducibility for the skip line measurements on US 421. The average reproducibility was ± 8.7 percent, almost double the reproducibility of the US 1 skip lines. Approximately 31 percent of all data points had a reproducibility of less than ± 5 percent, and about 76 percent of all data points had a reproducibility of less than ± 10 percent. The poor condition of the pavement markings may have contributed to the larger reproducibility values since small changes in the position of the measurements could result in large differences in readings.

White Edge Lines

The white edgeline retroreflectivity values at the US 1 test site ranged from 299 mcd/m^2/lux to 546 mcd/m^2/lux. These values were comparable to those obtained from the skip line measurements on US 1. Retroreflectivity values for the US 421 edgelines went from 21 mcd/m^2/lux to 273 mcd/m^2/lux. These values were generally higher than those measured for the skip lines on US 421. Figures 16 and 17 show the retroreflectivity values obtained at the two test sites. The units tested seemed to produce consistent results at both sites.

Figure 18 shows the reproducibility for the units at the US 1 test site. The units showed reproducibility comparable to that achieved for the same retroreflectivity values in the ambient light laboratory testing. The average reproducibility across all 88 data points was ±5.7 percent, which is comparable to that obtained on the US 1 skip lines. Reproducibility was ±5 percent or less at 32 percent of the data collection points, and was less than ±10 percent at 98 percent of the data points.

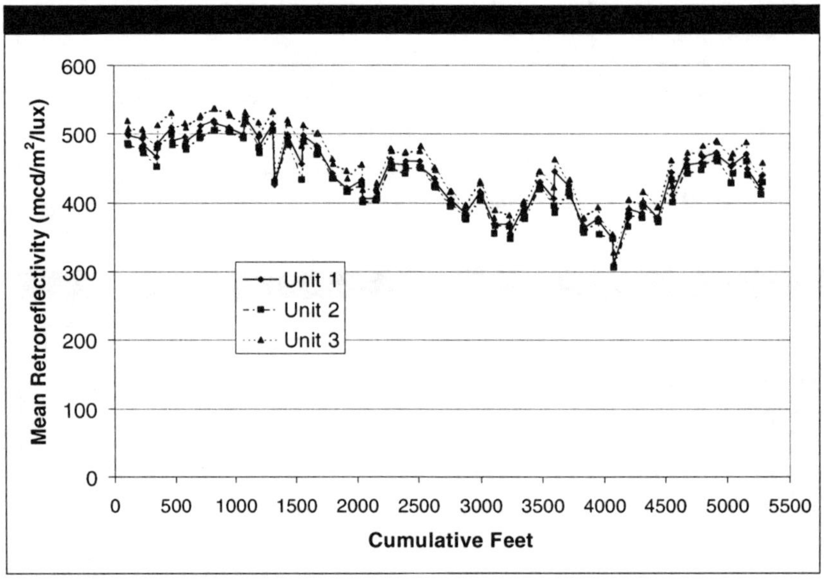

Figure 16. US 1 White Edgeline Retroreflectivity

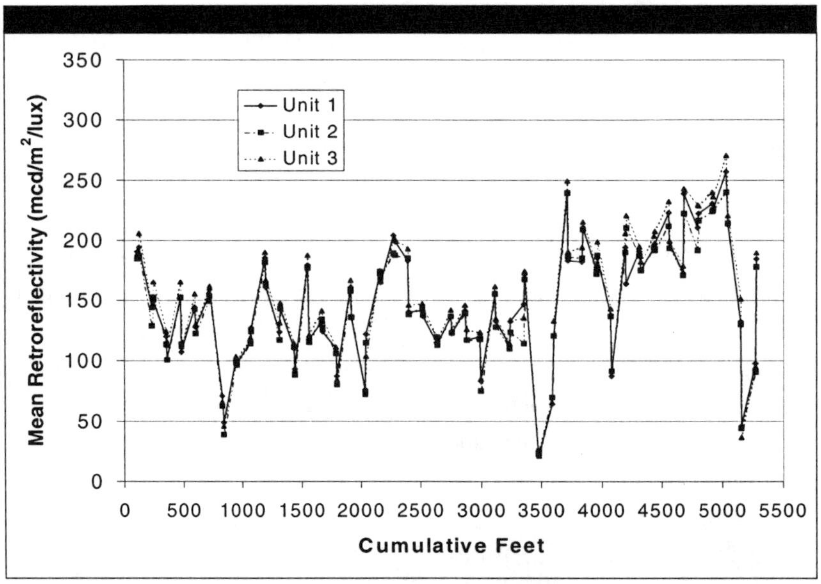

Figure 17. US 421 White Edgeline Retroreflectivity

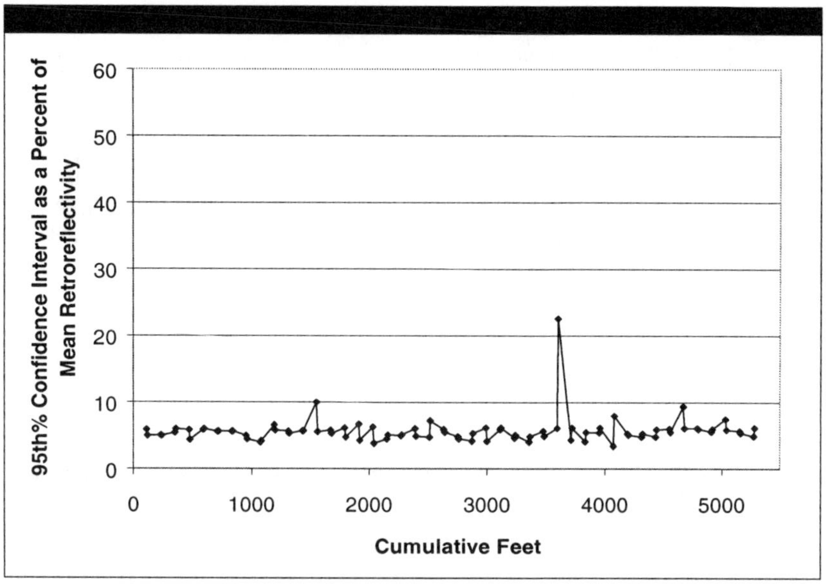

Figure 18. US 1 White Edgeline Reproducibility

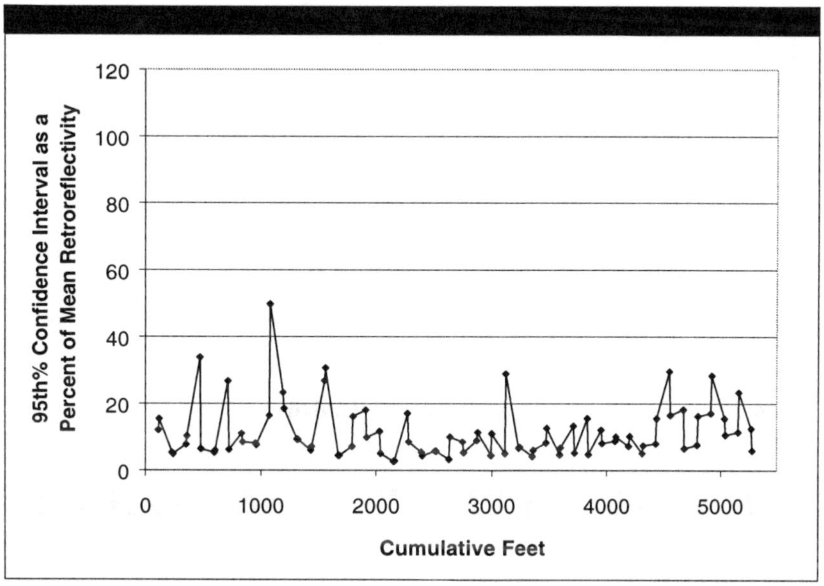

Figure 19. US 421 White Edgeline Reproducibility

Figure 19 shows the reproducibility for the white edge line measurements on US 421. Reproducibility was better on US 1 than on US 421. Approximately 15 percent of all data points had a reproducibility of less than ±5 percent, and about 43 percent of all data points had a reproducibility of less than ±10 percent. The average reproducibility across all 88 data points was ±11.4 percent. Some of the large reproducibility values are attributable to small changes in readings generating large percentages at low retroreflectivity values. Reproducibility appears to be better for the white edgeline than the skip lines for the lower retroreflectivity values.

Yellow Edgeline

The measured retroreflectivity values at both sites for the yellow edgeline tended to cover a smaller range of values than either the white skip lines or white edgeline. The yellow edgeline retroreflectivity values at the US 1 test site ranged from 101 mcd/m^2/lux to 310 mcd/m^2/lux. These values were significantly lower than those obtained from the white edgeline measurements on US 1. Retroreflectivity values for the US 421 edgelines went from 52 mcd/m²/lux to 102 mcd/m²/lux. These values were lower than those measured for the white edgelines on US 421. Figures 20 and 21 show the retroreflectivities of the yellow edgelines at the two sites. The three units tested appeared to produce similar results.

Figure 22 shows the reproducibility for the units at the US 1 test site. The overall average reproducibility was ±5.4 percent. Reproducibility was ±5 percent or less at 53 percent of the data collection points. About 97 percent of the data points had a reproducibility of less than ±10 percent.

Figure 23 shows the reproducibility for yellow edgeline measurements on US 421. Only 19 percent of all data points had a reproducibility of less than ±5 percent, but about 98 percent of all data points had a reproducibility of less than ±10 percent. The average reproducibility was ±6.2 percent. Some of the large reproducibility values are attributable to small changes in readings generating large percentages at low retroreflectivity values.

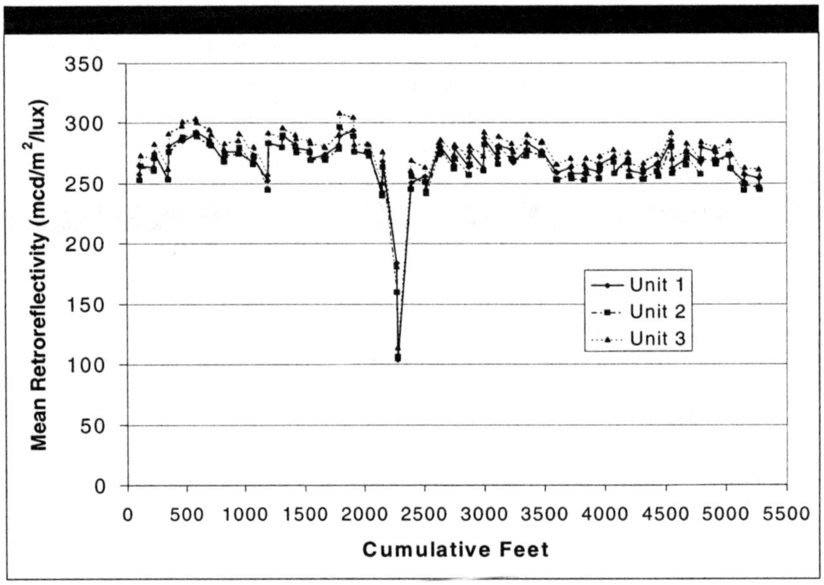

Figure 20. US 1 Yellow Edgeline Retroreflectivity

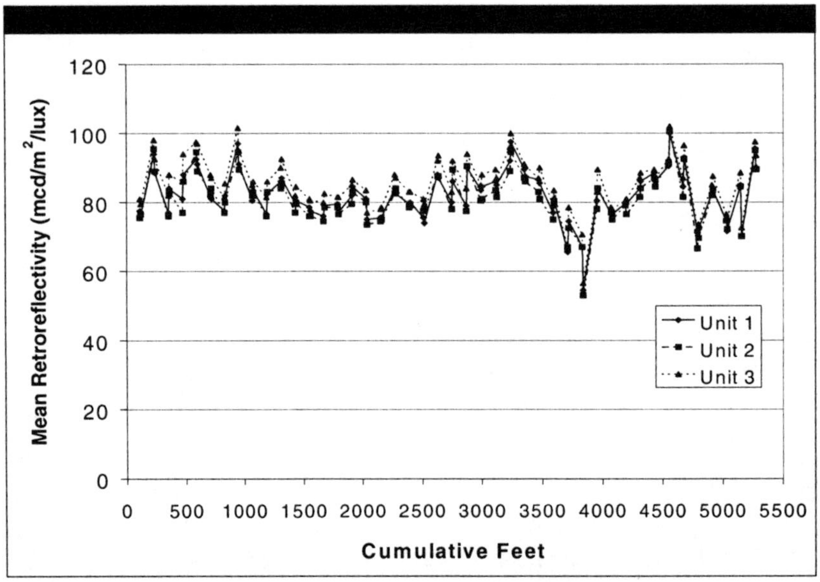

Figure 21. US 421 Yellow Edgeline Retroreflectivity

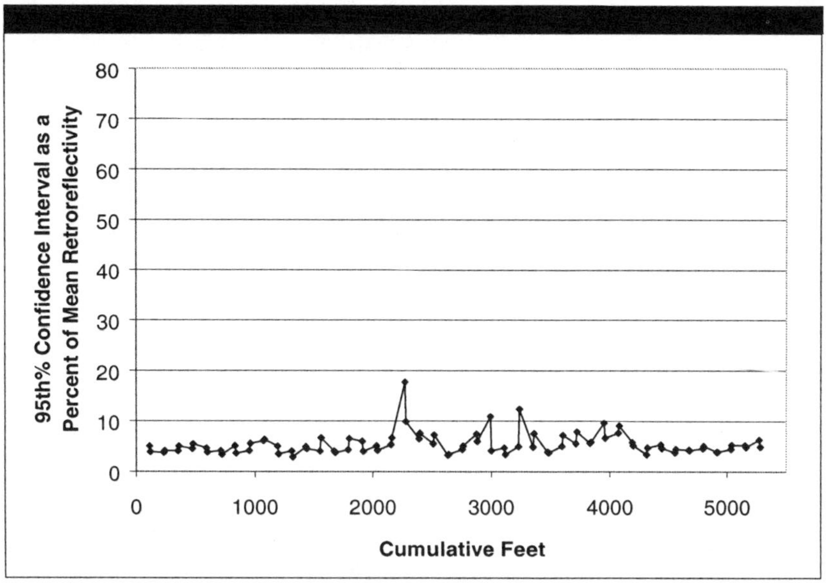

Figure 22. US 1 Yellow Edgeline Reproducibility

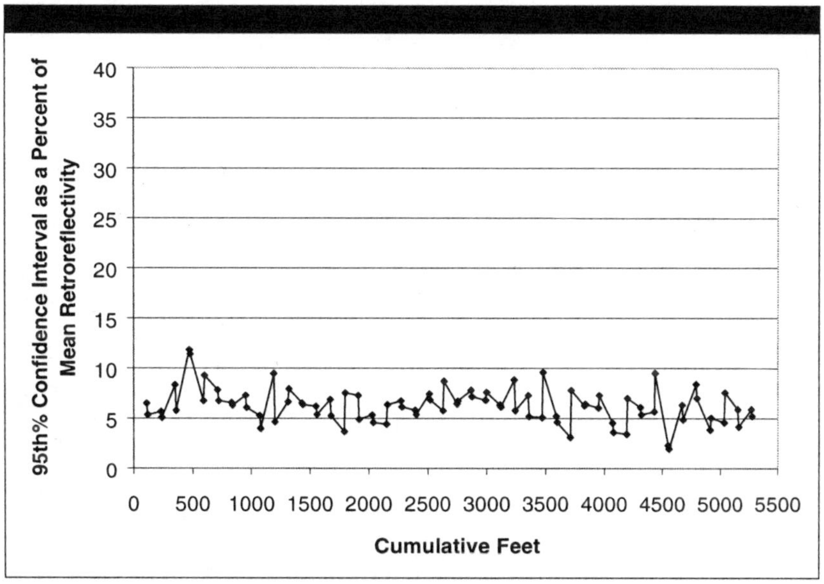

Figure 23. US 421 Yellow Edgeline Reproducibility

CHAPTER 5

Product Performance Summary

LTL 2000 Pavement Marking Retroreflectometer

The LTL 2000 measures pavement marking retroreflectivity according to European standard EN 1436 and ASTM E 1740. The unit utilizes the 30 meter geometry and measures pavement markings underneath the instrument that are shielded from ambient light.

The laboratory and field testing provided insight into the measurement bias, repeatability, and reproducibility of the unit for a variety of conditions. This chapter summarizes the results of the data collection effort.

5.1 Laboratory Panel Retroreflectivity

The average panel readings from the repeatability testing are summarized in Summary Table 1. The evaluation found that:

- The laboratory testing tended to produce retroreflectivity values that were less than the assigned values for panels B, C, D, G, and H. Measurements of panels E and F often produced values in excess of the assigned panel retroreflectivity.
- The nighttime, indoor tests produced lower retroreflectivity readings than the daytime and nighttime outdoor tests for panels B, D, E, and H.

Condition	Panel						
	B	C	D	E	F	G	H
High Temp., High Humidity	45.4	93.6	265.3	714.4	1230.8	175.7	71.4
High Temp., Low Humidity	46.0	102.6	278.7	714.3	1196.8	187.1	70.5
Low Temp., Low Humidity	47.1	94.8	281.8	663.0	1136.5	182.9	64.2
Daytime	46.4	96.9	273.9	707.3	1212.6	185.6	69.7
Night, Indoors	45.7	97.6	271.8	695.3	1182.6	186.3	68.5
Night, Outdoors	47.5	104.5	273.1	704.2	1168.4	186.8	69.0
Assigned Values	61.0	114.0	280.7	693.0	1133.7	195.0	85.3

Summary Table 1. Mean Panel Retroreflectivity (mcd/m^2/lux) For Repeatability Tests

5.2 Laboratory Panel Repeatability

The repeatability of the LTL 2000 is shown in Summary Table 2. The results indicate that:

- The daytime testing produced the largest repeatability results for every panel except B.
- The high temperature, high humidity testing produced larger repeatability results than all other environmental chamber testing, although several factors may have influenced these readings.

5.3 Laboratory Panel Reproducibility

The repeatability of the LTL 2000 was usually better than its reproducibility. Summary Table 3 shows the reproducibility results for the conditions tested. The results indicate that:

- The data collected outdoors at night had the best reproducibility for all panels except G.
- The high temperature, high humidity testing produced larger reproducibility values than other environmental chamber conditions for panels C, D, F, G, and H, although several factors may have influenced these values.
- The nighttime, indoors testing produced the larger reproducibility values than the other ambient light conditions for panels B, C, E, G, and H.

Condition	Panel						
	B	C	D	E	F	G	H
Low Temp., Low Humidity	±1.2%	±0.8%	±0.3%	±0.1%	±0.1%	±0.3%	±1.1%
High Temp., Low Humidity	0%	±0.6%	±0.6%	±0.1%	±0.1%	±0.3%	0%
High Temp., High Humidity	±8.1%	±5.7%	±1.2%	±0.4%	±0.4%	±2.2%	±4.7%
Daytime	±3.4%	±10.3%	±3.0%	±6.5%	±3.7%	±3.2%	±7.2%
Night, Indoors	±1.6%	±0.6%	±0.3%	±0.1%	±0.1%	±0.5%	0%
Night, Outdoors	0%	0%	±0.2%	±0.1%	±0.1%	±0.5%	0%
Average	±2.4%	±3.0%	±0.9%	±1.2%	±0.8%	±1.2%	±2.2%

Summary Table 2. Mean Repeatability For All Laboratory Tests (shown as 95th% Confidence Interval as a Percent of Mean Retroreflectivity*)

Condition	Panel						
	B	C	D	E	F	G	H
Low Temp., Low Humidity	±6.2%	±5.3%	±6.4%	±6.2%	±7.7%	±6.0%	±7.9%
High Temp., Low Humidity	±5.8%	±6.1%	±8.3%	±8.9%	±10.9%	±8.2%	±8.3%
High Temp., High Humidity	±5.7%	±7.8%	±8.5%	±6.8%	±11.7%	±23.2%	±11.9%
Daytime	±5.4%	±5.5%	±8.1%	±6.0%	±8.1%	±4.7%	±5.7%
Night, Indoors	±7.6%	±8.5%	±7.8%	±7.1%	±7.2%	±11.5%	±7.9%
Night, Outdoors	0	±2.1%	±3.7%	±3.7%	±5.9%	±5.5%	±5.5%
Average	±5.1%	±5.9%	±7.1%	±6.5%	±8.6%	±9.9%	±7.9%

Summary Table 3. Mean Reproducibility For All Laboratory Tests (shown as 95th% Confidence Interval as a Percent of Mean Retroreflectivity*)

***Lower percentages indicate the unit provides more consistent results than higher percentages.**

5.4 Shade Test

The shade test revealed that the readings taken in the shade were slightly larger than those taken in direct sunlight. However, these values were never more than 1 percent larger than the direct sunlight readings.

5.5 Field Data

The field data collection sites exhibited a wide range of retroreflectivity values. Average reproducibility was under ±12 percent for all sites. The three units appear to produce consistent results for all of the field sites. The US 421 consistently produced larger reproducibilities than the US 1 site. This is probably due the fact that slight variations in instrument placement could result in larger relative differences in readings on the less reflective, less uniform pavement markings. Summary Table 4 shows the results of the field testing.

Site	Type of Marking	Minimum ($mcd/m^2/lux$)	Mean ($mcd/m^2/lux$)	Maximum ($mcd/m^2/lux$)	Standard Deviation	Average Reproducibility*
US 1	Skip Line	381	466	561	37.7	±4.9%
US 1	White Edgeline	299	443	546	50.5	±5.7%
US 1	Yellow Edgeline	101	269	310	24.4	±5.4%
US 421	Skip Line	18	93	149	27.8	±8.7%
US 421	White Edgeline	21	148	273	50.1	±11.4%
US 421	Yellow Edgeline	52	83	102	7.8	±6.2%

Summary Table 4. Field Data Retroreflectivity Summary

***Shown as 95th% Confidence Interval as a Percent of Mean Retroreflectivity. Lower percentages indicate the unit provides more consistent results than higher percentages.**

APPENDIX A

Laboratory Data Summaries

Table A-1. Repeatability at High Temperature, High Humidity

Panel	Panel Set	Minimum R_L (mcd/m²/lux)	Maximum R_L (mcd/m²/lux)	Mean R_L (mcd/m²/lux)	Standard Deviation	95th% Confidence Interval
A	1	-1	-1	-1.0	0	*
	2	2	2	2.0	0	*
	3	-2	1	-0.5	1.2	-0.5 ± 2.4
B	1	47	47	47.0	0	*
	2	48	48	48.0	0	*
	3	36	44	41.3	3.2	41.3 ± 6.4
C	1	94	94	94.0	0	*
	2	98	99	98.3	0.5	98.3 ± 1.0
	3	80	92	88.5	4.6	88.5 ± 9.2
D	1	264	264	264.0	0	*
	2	259	260	259.8	0.4	259.8 ± 0.8
	3	269	277	272.2	2.6	272.2 ± 5.3
E	1	720	722	721.3	0.8	721.3 ± 1.6
	2	678	681	679.5	1.4	679.5 ± 2.8
	3	740	745	742.3	2.0	742.3 ± 3.9
F	1	1281	1285	1283.2	1.5	1283.2 ± 2.9
	2	1170	1172	1171.2	1.0	1171.2 ± 2.0
	3	1232	1242	1238.0	3.5	1238.0 ± 7.0
G	1	179	181	180.3	0.8	180.3 ± 1.6
	2	187	188	187.5	0.5	187.5 ± 1.1
	3	156	165	159.3	3.1	159.3 ± 6.3
H	1	65	65	65.0	0	*
	2	72	72	72.0	0	*
	3	73	80	77.2	2.9	77.2 ± 5.9

*** No confidence interval was computed since all measurements were identical**

Table A-2. Reproducibility for High Temperature, High Humidity

Panel	Panel Set	Minimum R_L (mcd/m²/lux)	Maximum R_L (mcd/m²/lux)	Mean R_L (mcd/m²/lux)	Standard Deviation	95th% Confidence Interval
A	1	-3	4	1.0	2.4	1.0 ± 4.7
	2	-1	2	0.3	1.0	0.3 ± 2.1
	3	-3	3	0.2	2.1	0.2 ± 4.3
B	1	45	48	47.0	1.1	47.0 ± 2.2
	2	45	48	46.7	1.0	46.7 ± 2.1
	3	40	45	43.0	1.7	43.0 ± 3.3
C	1	93	102	98.0	3.3	98.0 ± 6.6
	2	90	102	97.3	4.5	97.3 ± 8.9
	3	84	92	88.0	3.2	88.0 ± 6.4
D	1	252	280	268.7	11.8	268.7 ± 23.7
	2	246	279	263.8	13.6	263.8 ± 27.1
	3	235	251	241.7	5.9	241.7 ± 11.8
E	1	709	775	741.5	21.8	741.5 ± 43.7
	2	668	736	706.5	27.1	706.5 ± 54.1
	3	675	744	717.0	24.7	717.0 ± 49.3
F	1	1137	1304	1239.0	77.5	1239.0 ± 155.1
	2	1114	1333	1230.2	91.3	1230.2 ± 182.6
	3	1149	1240	1200.7	65.8	1200.7 ± 143.4
G	1	155	185	174.2	12.4	174.2 ± 24.9
	2	154	191	175.2	16.5	175.2 ± 32.9
	3	119	180	158.5	27.1	158.5 ± 54.2
H	1	57	67	64.0	3.6	64.0 ± 7.2
	2	68	74	70.7	2.3	70.7 ± 4.7
	3	68	81	76.3	5.8	76.3 ± 11.6

Table A-3. Repeatability at High Temperature, Low Humidity

Panel	Panel Set	Minimum R_L (mcd/m²/lux)	Maximum R_L (mcd/m²/lux)	Mean R_L (mcd/m²/lux)	Standard Deviation	95th% Confidence Interval
A	1	1	1	1.0	0	*
	2	2	3	2.3	0.5	2.3 ± 1.0
B	1	47	47	47.0	0	*
	2	45	45	45.0	0	*
C	1	103	104	103.2	0.4	103.2 ± 0.8
	2	102	102	102.0	0	*
D	1	270	279	275.0	0.5	275.5 ± 1.1
	2	281	322	295.5	1.0	283.0 ± 2.0
E	1	716	717	716.8	0.4	716.8 ± 0.8
	2	711	712	711.8	0.4	711.8 ± 0.8
F	1	1161	1163	1161.8	1.0	1161.8 ± 2.0
	2	1231	1232	1231.7	0.5	1231.7 ± 1.0
G	1	181	181	181.0	0	*
	2	193	194	193.2	0.2	193.2 ± 0.4
H	1	69	69	69.0	0	*
	2	72	72	72.0	0	*

*** No confidence interval was computed since all measurements were identical.**

Table A-4. Reproducibility for High Temperature, Low Humidity

Panel	Panel Set	Minimum R_L (mcd/m²/lux)	Maximum R_L (mcd/m²/lux)	Mean R_L (mcd/m²/lux)	Standard Deviation	95th% Confidence Interval
A	1	-1	2	0.3	1.3	0.3 ± 2.5
	2	2	3	2.5	0.6	2.5 ± 1.2
	3	-1	0	-0.5	0.6	-0.5 ± 1.2
B	1	46	47	46.5	0.6	46.5 ± 1.2
	2	45	50	46.8	2.2	46.8 ± 4.4
	3	44	44	44.0	0	*
C	1	99	101	100.3	1.0	100.3 ± 1.9
	2	103	113	105.5	5.0	105.5 ± 10.0
	3	93	96	94.3	1.5	94.3 ± 3.0
D	1	270	279	275.0	4.2	275.0 ± 8.5
	2	281	322	295.5	18.7	295.5 ± 37.3
	3	268	280	274.0	6.4	274.0 ± 12.8
E	1	681	730	697.5	22.7	697.5 ± 45.3
	2	710	816	747.5	48.6	747.5 ± 97.2
	3	707	735	721.8	15.3	721.8 ± 30.7
F	1	1126	1208	1163.5	40.7	1163.5 ± 81.5
	2	1240	1464	1325.8	104.2	1325.8 ± 208.3
	3	1196	1260	1228.5	33.7	1228.5 ± 67.3
G	1	173	187	179.0	6.7	179.0 ± 13.4
	2	195	219	204.3	10.7	204.3 ± 21.4
	3	174	181	177.5	4.0	177.5 ± 8.1
H	1	66	69	67.0	1.4	67.0 ± 2.8
	2	71	82	74.5	5.1	74.5 ± 10.1
	3	79	80	79.8	0.5	79.8 ± 1.0

Table A-5. Repeatability for Low Temperature, Low Humidity

Panel	Panel Set	Minimum R_L (mcd/m^2/lux)	Maximum R_L (mcd/m^2/lux)	Mean R_L (mcd/m^2/lux)	Standard Deviation	95th% Confidence Interval
A	1	0	0	0	0	*
	2	2	3	2.8	0.4	2.8 ± 0.8
B	1	47	47	47.0	0	*
	2	47	48	47.2	0.4	47.2 ± 0.8
C	1	92	92	92.0	0	*
	2	97	98	97.5	0.5	97.5 ± 1.1
D	1	293	294	293.8	0.4	293.8 ± 0.8
	2	269	270	269.8	0.4	269.8 ± 0.8
E	1	690	691	690.2	0.4	690.2 ± 0.8
	2	635	636	635.8	0.4	635.8 ± 0.8
F	1	1129	1130	1129.8	0.4	1129.0 ± 0.8
	2	1143	1144	1143.2	0.4	1143.2 ± 0.8
G	1	186	187	186.8	0.4	186.6 ± 0.8
	2	179	179	179.0	0	*
H	1	62	63	62.3	0.5	62.3 ± 1.0
	2	66	66	66.0	0	*

*** No confidence interval was computed since all measurements were identical.**

Table A-6. Reproducibility for Low Temperature, Low Humidity

Panel	Panel Set	Minimum R_L (mcd/m²/lux)	Maximum R_L (mcd/m²/lux)	Mean R_L (mcd/m²/lux)	Standard Deviation	95th% Confidence Interval
A	1	-1	1	0	1.2	0 ± 2.3
	2	1	3	1.8	1.0	1.8 ±1.9
B	1	47	47	47.0	0	*
	2	44	49	47.0	0	*
C	1	93	98	94.8	1.5	94.8 ± 3.0
	2	90	99	97.5	0.6	97.5 ± 1.2
D	1	278	293	289.5	4.0	289.5 ± 8.1
	2	254	288	274.5	4.7	274.5 ± 9.3
E	1	659	690	681.3	9.0	681.3 ± 18.0
	2	606	673	646.5	15.0	646.5 ± 30.0
F	1	1070	1236	1102.5	32.6	1102.5 ± 65.1
	2	1115	1229	1184.3	52.0	1184.3 ± 104.0
G	1	180	187	183.5	4.0	183.5 ± 8.1
	2	174	193	184.0	2.3	184.0 ± 4.6
H	1	62	70	62.5	0.6	62.5 ± 1.2
	2	65	80	67.8	2.2	67.8 ± 4.4

*** No confidence interval was calculated since all measurements were identical.**

Table A-7. Repeatability for Ambient Daylight

Panel	Panel Set	Minimum R_L (mcd/m^2/lux)	Maximum R_L (mcd/m^2/lux)	Mean R_L (mcd/m^2/lux)	Standard Deviation	95th% Confidence Interval
A	1	-2	2	0.4	1.7	0.4 ± 3.4
	2	2	4	3.1	1.0	3.1 ± 2.0
B	1	45	46	45.8	0.4	45.8 ± 0.8
	2	46	48	47.0	1.0	47.0 ± 2.1
C	1	87	100	93.5	6.7	93.5 ± 13.4
	2	98	103	100.2	2.3	100.2 ± 4.6
D	1	260	262	261.3	0.8	261.3 ± 1.6
	2	280	292	286.4	5.8	286.4 ± 11.6
E	1	695	757	726.1	32.3	726.1 ± 64.6
	2	684	694	688.6	4.8	688.6 ± 9.6
F	1	1162	1218	1190.0	28.7	1190.0 ± 57.5
	2	1222	1248	1235.3	13.0	1235.3 ± 26.0
G	1	182	189	185.5	3.7	185.5 ± 7.3
	2	183	188	185.7	2.0	185.7 ± 3.9
H	1	64	70	67.0	3.1	67.0 ± 6.3
	2	70	74	72.3	1.6	72.3 ± 3.2

Table A-8. Reproducibility for Ambient Daylight

Panel	Panel Set	Minimum R_L (mcd/m^2/lux)	Maximum R_L (mcd/m^2/lux)	Mean R_L (mcd/m^2/lux)	Standard Deviation	95th% Confidence Interval
A	1	3	6	4.3	1.5	4.3 ± 3.0
	2	-1	7	2.2	2.2	2.2 ± 4.5
	3	0	5	2.8	2.1	2.8 ± 4.1
B	1	45	47	46.3	1.0	46.3 ± 1.9
	2	44	48	46.6	1.3	46.6 ± 2.6
	3	42	45	43.4	1.2	43.4 ± 2.4
C	1	88	93	90.3	2.6	90.3 ± 5.3
	2	96	104	101.2	2.9	101.2 ± 5.9
	3	90	95	92.1	2.0	92.1 ± 4.1
D	1	260	267	263.5	3.5	263.5 ± 7.0
	2	258	293	279.3	13.6	279.3 ± 27.2
	3	264	288	275.1	8.8	275.1 ± 17.5
E	1	688	707	697.8	8.8	697.8 ± 17.5
	2	681	763	721.4	26.2	721.4 ± 52.3
	3	696	746	722.9	17.4	722.9 ± 34.9
F	1	1119	1191	1156.8	35.6	1156.8 ± 71.2
	2	1162	1337	1251.2	55.8	1251.2 ± 111.7
	3	1174	1298	1228.4	45.0	1228.4 ± 90.0
G	1	179	185	182.0	3.5	182.0 ± 6.9
	2	183	198	190.2	5.1	190.2 ± 10.2
	3	174	182	178.3	3.3	178.3 ± 6.6
H	1	65	66	65.3	0.5	65.3 ± 1.0
	2	69	76	72.2	2.6	72.2 ± 5.1
	3	74	79	76.4	1.6	76.4 ± 3.2

Table A-9. Repeatability for Night (Outside)

Panel	Panel Set	Minimum R_L (mcd/m^2/lux)	Maximum R_L (mcd/m^2/lux)	Mean R_L (mcd/m^2/lux)	Standard Deviation	95th% Confidence Interval
A	1	4	4	4.0	0	*
	2	2	2	2.0	0	*
B	1	48	48	48.0	0	*
	2	47	47	47.0	0	*
C	1	106	106	106.0	0	*
	2	103	103	103.0	0	*
D	1	261	262	261.2	0.4	261.2 ± 0.8
	2	285	285	285.0	0	*
E	1	736	737	736.3	0.5	736.3 ± 1.0
	2	672	672	672.0	0	*
F	1	1149	1150	1149.7	0.5	1149.7 ± 1.0
	2	1187	1188	1187.2	0.4	1187.2 ± 0.8
G	1	186	187	186.2	0.4	186.2 ± 0.8
	2	187	188	187.5	0.5	187.5 ± 1.1
H	1	67	67	67.0	0	*
	2	71	71	71.0	0	*

*** No confidence interval was computed since all measurements were identical.**

Table A-10. Reproducibility for Night (Outside)

Panel	Panel Set	Minimum R_L (mcd/m^2/lux)	Maximum R_L (mcd/m^2/lux)	Mean R_L (mcd/m^2/lux)	Standard Deviation	95th% Confidence Interval
A	1	4	5	4.3	0.5	4.3 ± 1.0
	2	1	3	1.8	1.0	1.8 ± 1.9
	3	3	5	4.3	1.0	4.3 ± 1.9
B	1	48	48	48.0	0	*
	2	47	47	47.0	0	*
	3	45	45	45.0	0	*
C	1	104	106	105.0	0.8	105.0 ± 1.6
	2	102	105	103.5	1.3	103.5 ± 2.6
	3	93	95	93.8	1.0	93.8 ± 1.9
D	1	254	264	259.0	4.8	259.0 ± 9.5
	2	279	291	284.0	6.0	284.0 ± 12.0
	3	271	279	275.0	4.1	275.0 ± 8.2
E	1	721	739	730.8	9.1	730.8 ± 18.2
	2	677	714	694.8	18.0	694.8 ± 36.1
	3	705	732	719.3	11.2	719.3 ± 22.4
F	1	1106	1163	1136.8	29.5	1136.8 ± 59.0
	2	1193	1276	1227.8	40.3	1227.8 ± 80.6
	3	1188	1254	1225.3	35.6	1225.3 ± 71.2
G	1	176	187	182.0	5.0	182.0 ± 9.9
	2	188	202	194.5	6.0	194.5 ± 11.9
	3	178	186	182.3	4.3	182.3 ± 8.7
H	1	66	69	67.3	1.3	67.3 ± 2.5
	2	74	80	76.3	2.6	76.3 ± 5.3
	3	78	82	79.8	2.1	79.8 ± 4.1

*** No confidence interval was computed since all measurements were identical.**

Table A-11. Repeatability for Night (Inside)

Panel	Panel Set	Minimum R_L (mcd/m²/lux)	Maximum R_L (mcd/m²/lux)	Mean R_L (mcd/m²/lux)	Standard Deviation	95th% Confidence Interval
A	1	2	2	2.0	0	*
	2	2	2	2.0	0	*
B	1	45	46	45.3	0.5	45.3 ± 1.0
	2	46	46	46.0	0	*
C	1	95	95	95.0	0	*
	2	100	101	100.2	0.4	100.2 ± 0.8
D	1	264	265	264.5	0.5	264.5 ± 1.1
	2	279	279	279.0	0	*
E	1	718	719	718.2	0.4	718.2 ± 0.8
	2	672	673	672.5	0.5	672.5 ± 1.1
F	1	1152	1153	1152.8	0.4	1152.8 ± 0.8
	2	1212	1213	1212.3	0.5	1212.3 ± 1.0
G	1	185	186	185.2	0.4	185.2 ± 0.8
	2	187	188	187.3	0.5	187.3 ± 1.0
H	1	67	67	67.0	0	*
	2	70	70	70.0	0	*

*** No confidence interval was computed since all measurements were identical.**

Table A-12. Reproducibility for Night (Inside)

Panel	Panel Set	Minimum R_L (mcd/m^2/lux)	Maximum R_L (mcd/m^2/lux)	Mean R_L (mcd/m^2/lux)	Standard Deviation	95th% Confidence Interval
A	1	1	3	2.3	1.0	2.3 ± 1.9
	2	1	3	2.3	1.0	2.3 ± 1.9
	3	1	3	2.0	0.8	2.0 ± 1.6
B	1	45	46	45.8	0.5	45.8 ± 1.0
	2	46	52	47.8	2.9	47.8 ± 5.7
	3	42	43	42.5	0.6	42.5 ± 1.2
C	1	93	97	95.3	2.1	95.3 ± 4.1
	2	96	110	100.3	6.6	100.3 ± 13.1
	3	91	95	92.8	1.7	92.8 ± 3.4
D	1	257	266	261.3	4.9	261.3 ± 9.8
	2	280	318	294.0	16.6	294.0 ± 33.1
	3	261	274	267.8	6.7	267.8 ± 13.4
E	1	698	723	711.0	13.3	711.0 ± 26.7
	2	667	758	702.5	38.9	702.5 ± 77.8
	3	700	725	712.0	13.9	712.0 ± 27.8
F	1	1095	1162	1128.0	37.5	1128.0 ± 75.1
	2	1215	1366	1289.0	61.7	1289.0 ± 123.3
	3	1174	1213	1193.5	21.4	1193.5 ± 42.8
G	1	156	187	175.5	13.5	175.5 ± 27.0
	2	187	214	198.0	11.4	198.0 ± 22.8
	3	170	178	173.5	4.1	173.5 ± 8.2
H	1	66	68	67.0	1.2	67.0 ± 2.3
	2	68	78	71.8	4.3	71.8 ± 8.7
	3	73	77	75.0	1.8	75.0 ± 3.7

APPENDIX B

Field Data Summaries

Table B-1. US 1 White Edgeline Retroreflectivity Aggregated at 480 ft Intervals

Distance (ft)	Unit	Minimum (mcd/m²/lux)	Maximum (mcd/m²/lux)	Mean (mcd/m²/lux)	Standard Deviation
0-480	1	463	510	490.9	12.7
	2	452	501	479.9	13.2
	3	480	537	508.0	14.5
480-960	1	485	523	506.6	10.4
	2	477	508	495.9	10.5
	3	509	542	525.9	9.8
960-1440	1	416	526	492.0	29.9
	2	417	519	484.2	26.1
	3	429	546	508.4	31.2
1440-1920	1	418	503	454.7	28.8
	2	415	492	445.9	27.0
	3	433	525	474.9	28.3
1920-2400	1	406	466	436.6	24.0
	2	400	457	429.3	22.5
	3	415	483	452.9	25.8
2400-2880	1	381	460	419.1	28.3
	2	371	454	411.8	29.0
	3	389	483	435.1	31.2
2880-3360	1	355	418	383.0	21.8
	2	347	412	376.4	21.7
	3	367	434	396.9	22.2
3360-3840	1	363	445	409.4	29.1
	2	334	438	395.8	31.6
	3	375	464	423.9	30.0
3840-4320	1	299	406	370.1	28.4
	2	302	399	362.9	26.7
	3	326	420	382.1	28.0
4320-4800	1	375	474	431.7	36.3
	2	370	460	422.4	33.5
	3	387	488	444.4	37.0
4800-5280	1	416	478	453.4	17.9
	2	407	467	441.9	18.2
	3	431	491	469.5	18.8

Table B-2. US 1 Yellow Edgeline Retroreflectivity Aggregated at 480 ft Intervals

Distance (ft)	Unit	Minimum (mcd/m²/lux)	Maximum (mcd/m²/lux)	Mean (mcd/m²/lux)	Standard Deviation
0-480	1	250	291	270.9	12.7
	2	250	290	269.3	13.4
	3	262	302	280.8	14.6
480-960	1	263	293	281.2	8.6
	2	264	293	280.1	7.9
	3	277	307	291.3	8.5
960-1440	1	251	293	275.5	12.1
	2	242	290	274.1	14.0
	3	255	298	283.3	13.1
1440-1920	1	264	300	279.6	9.6
	2	263	298	278.3	9.9
	3	279	310	289.4	10.9
1920-2400	1	101	281	230.5	57.7
	2	104	281	227.4	59.1
	3	114	287	239.5	58.6
2400-2880	1	238	281	268.0	13.2
	2	234	276	262.3	12.4
	3	245	288	274.2	12.2
2880-3360	1	250	292	275.9	11.6
	2	248	286	272.3	10.6
	3	261	299	283.1	10.7
3360-3840	1	250	278	263.9	9.1
	2	246	276	259.5	10.0
	3	256	286	271.6	9.3
3840-4320	1	248	280	262.9	9.1
	2	244	275	259.1	9.0
	3	256	287	270.4	7.9
4320-4800	1	261	286	271.4	8.3
	2	253	282	264.8	8.5
	3	266	293	277.3	8.5
4800-5280	1	243	279	260.9	10.9
	2	240	274	256.9	11.9
	3	255	285	269.3	10.7

Table B-3. US 1 White Skipline Retroreflectivity Aggregated at 480 ft Intervals

Distance (ft)	Unit	Minimum (mcd/m²/lux)	Maximum (mcd/m²/lux)	Mean (mcd/m²/lux)	Standard Deviation
0-480	1	475	546	511.8	21.0
	2	468	528	499.8	20.2
	3	493	561	525.8	22.2
480-960	1	422	525	496.7	25.2
	2	461	515	492.1	16.0
	3	488	546	517.4	17.2
960-1440	1	468	519	487.2	17.3
	2	457	500	478.4	14.4
	3	481	523	498.5	14.2
1440-1920	1	464	524	494.3	19.6
	2	456	510	481.4	18.3
	3	478	538	505.9	18.6
1920-2400	1	422	505	478.6	25.6
	2	412	490	465.7	23.1
	3	429	523	489.8	27.2
2400-2880	1	412	486	452.8	27.0
	2	405	474	440.4	25.3
	3	428	500	462.8	27.4
2880-3360	1	414	509	458.1	28.6
	2	402	489	444.8	25.9
	3	421	522	466.8	30.8
3360-3840	1	394	454	426.9	20.7
	2	384	464	417.4	22.9
	3	400	464	437.4	22.7
3840-4320	1	405	464	428.3	17.3
	2	392	453	417.9	18.0
	3	411	479	439.2	19.6
4320-4800	1	398	476	436.4	24.0
	2	381	451	421.3	22.3
	3	404	486	446.3	26.0
4800-5280	1	401	499	453.7	26.6
	2	389	483	440.6	25.2
	3	417	509	464.5	26.8

Table B-4. US 421 White Edgeline Retroreflectivity Aggregated at 480 ft Intervals

Distance (ft)	Unit	Minimum (mcd/m²/lux)	Maximum (mcd/m²/lux)	Mean (mcd/m²/lux)	Standard Deviation
0-480	1	101	196	145.0	33.7
	2	97	191	141.8	32.0
	3	107	208	152.3	33.8
480-960	1	45	159	112.9	38.8
	2	37	155	108.1	40.5
	3	39	165	115.1	42.3
960-1440	1	93	186	132.3	28.5
	2	88	183	130.5	29.6
	3	92	191	136.1	30.0
1440-1920	1	87	180	131.1	27.8
	2	80	177	129.3	28.9
	3	84	188	135.4	30.8
1920-2400	1	75	206	157.4	41.9
	2	69	193	154.0	40.6
	3	72	203	156.9	45.6
2400-2880	1	112	144	127.9	11.7
	2	112	146	128.8	11.7
	3	116	150	133.2	12.2
2880-3360	1	83	174	131.7	26.2
	2	74	168	124.0	27.8
	3	84	177	132.8	26.5
3360-3840	1	21	240	130.8	82.5
	2	21	243	132.3	81.9
	3	21	249	136.3	86.8
3840-4320	1	86	216	163.6	39.7
	2	86	215	168.7	36.6
	3	86	221	176.9	39.9
4320-4800	1	171	243	206.9	20.7
	2	168	230	199.0	17.2
	3	177	246	215.1	21.1
4800-5280	1	42	261	173.4	71.8
	2	43	240	168.1	69.8
	3	36	273	179.9	77.7

Table B-5. US 421 Yellow Edgeline Retroreflectivity Aggregated at 480 ft Intervals

Distance (ft)	Unit	Minimum (mcd/m²/lux)	Maximum (mcd/m²/lux)	Mean (mcd/m²/lux)	Standard Deviation
0-480	1	75	96	83.3	6.4
	2	74	97	82.3	7.3
	3	78	100	87.6	7.0
480-960	1	77	97	86.7	6.8
	2	75	97	86.4	6.6
	3	81	102	91.6	7.0
960-1440	1	75	88	82.0	3.5
	2	75	86	81.2	3.4
	3	78	94	85.9	3.9
1440-1920	1	75	85	79.3	3.0
	2	74	85	77.8	2.8
	3	80	87	82.3	2.4
1920-2400	1	74	86	79.1	3.7
	2	73	84	78.4	3.8
	3	76	88	82.3	4.1
2400-2880	1	74	94	82.4	6.3
	2	76	91	83.3	5.8
	3	79	95	87.4	5.9
2880-3360	1	82	98	88.2	4.7
	2	78	95	85.4	4.9
	3	84	102	90.9	5.0
3360-3840	1	54	87	73.9	10.6
	2	52	84	72.3	9.6
	3	56	91	76.4	10.9
3840-4320	1	75	87	80.8	3.6
	2	75	84	79.4	3.4
	3	77	90	83.3	4.5
4320-4800	1	70	100	85.7	9.7
	2	65	101	84.2	11.3
	3	71	102	87.5	10.2
4800-5280	1	69	98	81.6	8.8
	2	70	97	81.4	8.5
	3	72	100	84.6	8.8

Table B-6. US 421 White Skipline Retroreflectivity Aggregated at 480 ft Intervals

Distance (ft)	Unit	Minimum (mcd/m²/lux)	Maximum (mcd/m²/lux)	Mean (mcd/m²/lux)	Standard Deviation
0-480	1	49	138	101.3	22.4
	2	55	141	103.7	20.4
	3	63	146	105.0	20.8
480-960	1	75	118	95.1	13.0
	2	65	116	94.2	16.0
	3	69	124	98.8	14.4
960-1440	1	40	115	77.0	24.4
	2	41	103	78.2	21.8
	3	44	108	77.7	21.6
1440-1920	1	49	123	93.4	24.4
	2	54	124	95.1	24.0
	3	67	129	103.3	20.8
1920-2400	1	84	145	115.3	17.5
	2	97	144	116.3	14.4
	3	105	147	120.9	15.4
2400-2880	1	78	136	107.3	18.6
	2	79	139	108.7	18.1
	3	82	146	111.1	21.9
2880-3360	1	107	138	117.1	9.5
	2	82	146	117.2	15.1
	3	112	149	123.5	11.1
3360-3840	1	55	117	84.2	21.6
	2	61	113	84.1	19.1
	3	62	120	87.9	20.9
3840-4320	1	86	116	103.9	8.1
	2	79	115	101.3	9.9
	3	89	125	108.4	10.5
4320-4800	1	39	99	71.6	18.9
	2	48	104	71.8	19.9
	3	48	103	74.8	18.6
4800-5280	1	18	62	40.8	12.5
	2	18	62	43.2	12.7
	3	20	54	41.3	10.5

APPENDIX C

Photometric Range Testing of Panels

The photometric range testing was used in order to establish an "assigned value" for the panels. The "assigned" value should not be taken as a known or absolute value of the panel. The photometric ranges did produce measurements that varied, and it is impossible to determine which range gave the most accurate results since there is no accepted calibration standard. Table C-1 summarizes the results of the photometric range testing, including variations between ranges.

Method 1: Measurements conducted in a photometric laboratory at a 15 meter distance using the direct method as described in ASTM D 4061. This was done using facilities at 3M.
Method 2: Measurements obtained in the same 3M photometric laboratory at a 5 meter distance using the direct method in ASTM D 4061.
Method 3: Readings collected by 3M using the luminance measurement method (a telephotometer and a lux meter at a 5 meter distance).
Method 4: Measurements obtained at the Turner Fairbank Highway Research Center using the luminance measurement method. A different white calibration standard was used for each of the three trials.

Table C-1. Photometric Range Readings

Panel	Panel Set	Photometric Range Method 1		Photometric Range Method 2		Photometric Range Method 3		Photometric Range Method 4		Average of Methods	Range of Methods
		Average n=3	St. Dev.	Average n=3	St. Dev.	Average n=2	St. Dev.	Average n=3	St. Dev.	mcd/m²/lx	mcd/m²/lx
B	1	50	1.3	62	10.1	69	10.7	58	7.2	60	50 - 69
	2	54	3.1	79	8.9	46	10.5	65	9.5	61	46 - 79
	3	51	3.8	78	17.4	55	12.2	65	14.6	62	51 - 78
C	1	105	6.8	106	15.1	130	40.3	125	20.3	116	105 -130
	2	101	7.9	137	9.5	103	19.9	120	12.5	115	101 -137
	3	100	5.6	121	16.3	96	5.0	126	16.5	111	96 - 126
D	1	292	18.5	244	16.9	258	19.1	301	19.2	274	244 -301
	2	294	29.6	286	13.5	261	28.3	311	4.4	288	261 -311
	3	298	23.5	268	33.0	247	33.9	308	9.8	280	247 - 308
E	1	794	40.0	631	9.0	675	58.0	726	18.7	707	631 -794
	2	781	27.7	609	26.3	670	72.8	712	10.7	693	609 - 781
	3	739	22.5	598	42.3	672	70.7	708	8.3	679	598 - 739
F	1	1348	67.3	965	36.5	1053	102.5	1202	45.0	1142	965 - 1348
	2	1321	43.1	966	35.2	1060	118.1	1190	34.1	1134	966 - 1321
	3	1235	28.5	978	75.2	1059	113.8	1229	62.5	1125	978 - 1235
G	1	182	20.6	180	21.8			215	19.7	192	180 - 215
	2	202	0.5	194	21.7			191	6.8	195	191 -202
	3	196	12.3	188	24.6			211	18.4	198	188 - 211
H	1	69	2.0	78	21.2			83	16.5	77	69 - 83
	2	74	4.3	89	33.9			100	9.2	88	74 - 100
	3	76	2.4	97	8.0			99	16.6	91	76 - 99

APPENDIX D

Examples of Statistical Computations

Overview of Terms

The purpose of this appendix is to provide some background on the statistical analysis procedure used in this report. Two important statistical concepts were used in this evaluation: standard deviation and confidence interval. Standard deviation is defined by the equation:

$$\sigma = \sqrt{\frac{1}{n-1}\sum_{i=1}^{n}(x_i - \overline{x})^2}$$

Where:

σ = standard deviation
n = number of observations
$\overline{x}$ = average of observations
x_i = observation i

The standard deviation is a measure of the variability of a sample from the mean. *On the average*, samples will be within plus or minus one standard deviation of the mean, assuming there is a normal distribution of data. There will be data points that will be outside one standard deviation of the mean. Generally speaking, about ½ of the observations will usually fall within plus or minus one standard deviation of the mean.

Confidence interval is defined by the equation:

$$\overline{x} \pm z\left(\frac{\alpha}{2}\right)\frac{\sigma}{\sqrt{n}}$$

Where:

$\overline{x}$ = average of observations
$z(\alpha/2)$ = 2.0 for 95th percentile confidence interval (based on normal distribution of data)
σ = standard deviation
n = number of observations

The confidence interval expresses a range that covers the expected values of the mean with a certain probability. In this evaluation, a 95 percent probability was selected, meaning that an interval was computed that should include 95 percent of all readings taken with an instrument.

Analysis Procedure

A field user of the instrument would be most interested in what range of values they could expect the unit to produce if they took a single reading. That is, how much variation could be expected provided only one reading was collected. When computing the 95th percentile confidence interval, we operated under the assumption that there would be only one observation (n=1). Thus, the 95th percentile confidence interval is represented by the mean plus or minus approximately 2 standard deviations. This means that if a user takes a single reading with the instrument, 95% of the time it will fall into the range defined by the mean ±2 standard deviations.

In this evaluation, three distinct panel sets were tested during the repeatability and reproducibility portions of the testing. While panel sets 1, 2, and 3 were very similar, there were detectable differences in their retroreflectivity. In order to compute an overall repeatability or reproducibility value for a panel, the standard deviations needed to be computed for each individual panel set, and then a pooled standard deviation was calculated. The pooled standard deviation combines the standard deviation estimates from each panel set to create an overall standard deviation. The assumption is that the distribution of each of the panel sets is the same. The pooled standard deviation is defined as:

$$\sigma = \sqrt{\frac{(n_1 - 1)\sigma_1^2 + (n_2 - 1)\sigma_2^2}{n_1 + n_2 - 2}}$$

Where: n_i = Number of observations for panel set I
σ_i = standard deviation for panel set i

Repeatability and reproducibility percentages were computed using the confidence intervals resulting from the pooled standard deviations. The percentage was computed using the following equation:

$$\frac{2\sigma}{\bar{x}}$$

Where: σ = Pooled standard deviation
$\bar{x}$ = Overall mean

Examples of the calculation of repeatability and reproducibility percentages are included in this appendix to better illustrate this analysis procedure.

Repeatability Calculation Example

Data

Instrument	Panel Set	R_L (mcd/m²/lux)
1	1	115
1	1	115
1	1	114
1	1	117
1	1	112
1	1	114
2	2	110
2	2	110
2	2	110
2	2	110
2	2	109
2	2	110

Panel Set 1 Mean: 114.5
Panel Set 2 Mean: 109.8
Combined Mean: 112.2

Standard Deviation for Panel Set 1:

$$\sqrt{\frac{1}{n-1}\sum_{i=1}^{n}(x_i - \bar{x})^2}$$

$$\sqrt{\frac{1}{(6-1)}\sum(x_i - 114.5)^2}$$

$= 1.643$

Standard Deviation for Panel Set 2: 0.408

Pooled Standard Deviation

$$\sigma = \sqrt{\frac{(n_1-1)\sigma_1^2 + (n_2-1)\sigma_2^2}{n_1+n_2-2}}$$

$$\sigma = \sqrt{\frac{(6-1)1.643^2 + (6-1)0.408^2}{6+6-2}}$$

$$\sigma = 1.197$$

Combined 95th% Confidence Interval:

$$\left[\bar{x} - z\left(\frac{\alpha}{2}\right)\frac{\sigma}{\sqrt{n}}, \bar{x} + z\left(\frac{\alpha}{2}\right)\frac{\sigma}{\sqrt{n}}\right]$$, where $z(\alpha/2) = 2.0$ at 95th% confidence interval

$$\left[112.2 - 2.0\frac{1.197}{\sqrt{1}}, 112.2 + 2.0\frac{1.197}{\sqrt{1}}\right]$$

$= [112.2 \pm 2.4]$

The repeatability percentage would be $\pm$ 2.4/112.2 = $\pm$ 2.1%

Reproducibility Calculation Example

Data

Instrument	Panel Set	R_L (mcd/m²/lux)
1	1	1109
1	1	1130
2	1	1013
2	1	1024
1	2	1261
1	2	1247
2	2	1146
2	2	1130
1	3	1228
1	3	1208
2	3	1203
2	3	1179

Panel Set 1 mean: 1069.0
Panel Set 2 mean: 1196.0
Panel Set 3 mean: 1204.5
Combined Mean: 1156.5

Standard Deviation for Panel Set 1:

$$\sqrt{\frac{1}{n-1}\sum_{i=1}^{n}(x_i - \bar{x})^2}$$

$$\sqrt{\frac{1}{(4-1)}\sum(x_i - 1069)^2}$$

$= 59.11$

Standard Deviation for Panel Set 2: 67.53

Standard Deviation for Panel Set 3: 20.14

Pooled Standard Deviation

$$\sigma = \sqrt{\frac{(n_1-1)\sigma_1^2 + (n_2-1)\sigma_2^2 + (n_3-1)\sigma_3^2}{n_1+n_2+n_3-3}}$$

$$\sigma = \sqrt{\frac{(4-1)59.11^2 + (4-1)67.53^2 + (4-1)20.14^2}{4+4+4-3}}$$

$$\sigma = 53.10$$

Combined 95th% Confidence Interval:

$$\left[\bar{x} - z\left(\frac{\alpha}{2}\right)\frac{\sigma}{\sqrt{n}}, \bar{x} + z\left(\frac{\alpha}{2}\right)\frac{\sigma}{\sqrt{n}}\right]$$, where $z(\alpha/2) = 2.0$ at 95th% confidence interval

$$\left[1156.5 - 2.0\frac{53.1}{\sqrt{1}}, 1156.5 + 2.0\frac{53.1}{\sqrt{1}}\right]$$

$= [1156.5 \pm 106.2]$

The reproducibility percentage would be $\pm 106.2/1156.5 = \pm 9.2\%$

APPENDIX E

Modifications Performed After Testing

This appendix allows for vendors to provide information on modifications that they have made to their product after the HITEC laboratory and field testing was completed. In many cases, these modifications came as a direct result of instrument performance during testing. It should be emphasized that the information in this appendix was provided by the manufacturer. **This information has not been independently verified, and HITEC cannot vouch for the accuracy of this information.**

As noted in the report, at the beginning of the high temperature, high humidity testing, one of the LTL 2000 instruments displayed a "High Zero Warning" error during calibration. The purpose of an error message is to alert the user to a problem that may cause measurements to be incorrect. This error message was generated by the error warning system of the LTL 2000 and was caused by the drastic difference between the ambient office conditions and the high humidity conditions within the test chamber.

The susceptibility to humidity was traced to an assembly glitch on one of the printed circuit boards, which was corrected immediately after the conclusion of the laboratory testing. The error warning system provides assurance to the user that the LTL 2000 is functioning properly by monitoring all important aspects of the instrument, including battery charge, zero set, calibration, stray light, etc.

RECENT CERF PUBLICATIONS

A Nationwide Survey of Civil Engineering-Related R&D (#62970)
Provides the first comprehensive data on the current level of civil engineering-related R&D investment, funding organizations, research performers, and areas of research emphasis. *$50.00 ($37.50 for ASCE and CAB members)*

Accelerating Environmental Technology Implementation: A Survey of Present Practices and New Directions (#40288)
Details the results of the comprehensive EvTEC survey assessing environmental industry status and needs. *$35.00 ($25.00 for ASCE and CAB members)*

Action Plan: An Enhanced Building Technology Evaluation Process (#40198)
Presents the Action Plan for an enhanced national evaluation process for new building technologies, as developed by the Partnership for Building Innovation (PBI). *$20.00 ($15.00 for ASCE and CAB members)*

Commercializing Infrastructure Technologies: A Handbook for Innovators (#40134)
Written for the seasoned professional as well as the layperson, this handbook details real world methods and practices for moving innovative technologies from the drawing board into the marketplace. *$40.00 ($30.00 for ASCE and CAB members)*

Constructed Civil Infrastructure Systems R&D: A European Perspective (#40001)
Examines Constructed Civil Infrastructure Systems (CCIS) R&D in the United Kingdom, Sweden, Germany, the Netherlands, France and Italy. *$25.00 ($18.75 for ASCE and CAB members)*

Construction Industry Research Prospectuses for the 21st Century (#40186)
Describes the 38 prospectuses generated at CERF's recent *International Research Symposium: Engineering and Construction for Sustainable Development in the 21st Century*. *$75.00 ($56.25 for ASCE and CAB members)*

Creating the 21st Century through Innovation (#40185)
Executive summary of *Construction Industry Research Prospectuses for the 21st Century* (#40186). *$25.00 ($18.75 for ASCE and CAB members)*

Engineering and Construction for Sustainable Development in the 21st Century: Assessing Global Research Needs (#40141)
Identifies key engineering research and construction issues for the 21st century that support sustainable development. *$20.00 ($15.00 for ASCE and CAB members)*

Federal Public Works Infrastructure R&D: A New Perspective (#62943)
Provides an analysis of the federal role in public works infrastructure research and development. *$40.00 ($30.00 for ASCE and CAB members)*

Geo-Engineering: A Vision for the 21st Century (#40042)
Documents recent actions by U.S. experts in the geo-engineering related disciplines of civil engineering to define the role of geo-engineers and geo-engineering in the 21st century. *$25.00 ($18.75 for ASCE and CAB members)*

International Sourcebook for Construction Industry Product Assessment (#40173)
Presents a summary of the key characteristics of leading construction industry technical approval and evaluation organizations in the world. *$20.00 ($15.00 for ASCE and CAB members)*

Materials for Tomorrow's Infrastructure: A Ten-Year Plan for Deploying High-Performance Construction Materials and Systems—Executive Report (#40066)
Presents a high-level, policy-oriented summary of the High-Performance CONstruction MATerials (CONMAT) implementation plan. *$40.00 ($30.00 for ASCE and CAB members)*

Materials for Tomorrow's Infrastructure: A Ten-Year Plan for Deploying High-Performance Construction Materials and Systems—Technical Report (#40059)
75.00 ($56.25 for ASCE and CAB members)

Opportunities in Asia: An Assessment of Construction Trends, Needs, and Potential Collaborations (#40257)
This comprehensive report details CERF's 1996 Asia Technology Assessment Mission and its findings; includes detailed site visit reports. *$75.00 ($56.25 for ASCE and CAB members)*

The Foundation for a New Approach to Implement Building Innovation (#40126)
Focuses on the development of a proposed national strategy for public-private sector partnerships in implementing technology and systems innovation in the buildings sector. *$20.00 ($15.00 for ASCE and CAB members)*

Solving the Innovation Puzzle: Challenges Facing the U.S. Design and Construction Industry (#40023)
Co-authored by CERF president, this book outlines the obstacles to bringing innovative ideas into practice, including the influence of government policies and private-sector priorities. *$22.00 ($16.50 for ASCE and CAB members)*

Transferring Research Into Practice: Lessons from Japan's Construction Industry (#5505)
Documents the role of research in Japan's construction industry and provides insights into how the results of Japanese research activities enter practice so quickly. *$25.00 ($18.75 for ASCE and CAB members)*

To order, please call 800-548-2723 or 703-295-6300, or email pubs@asce.org